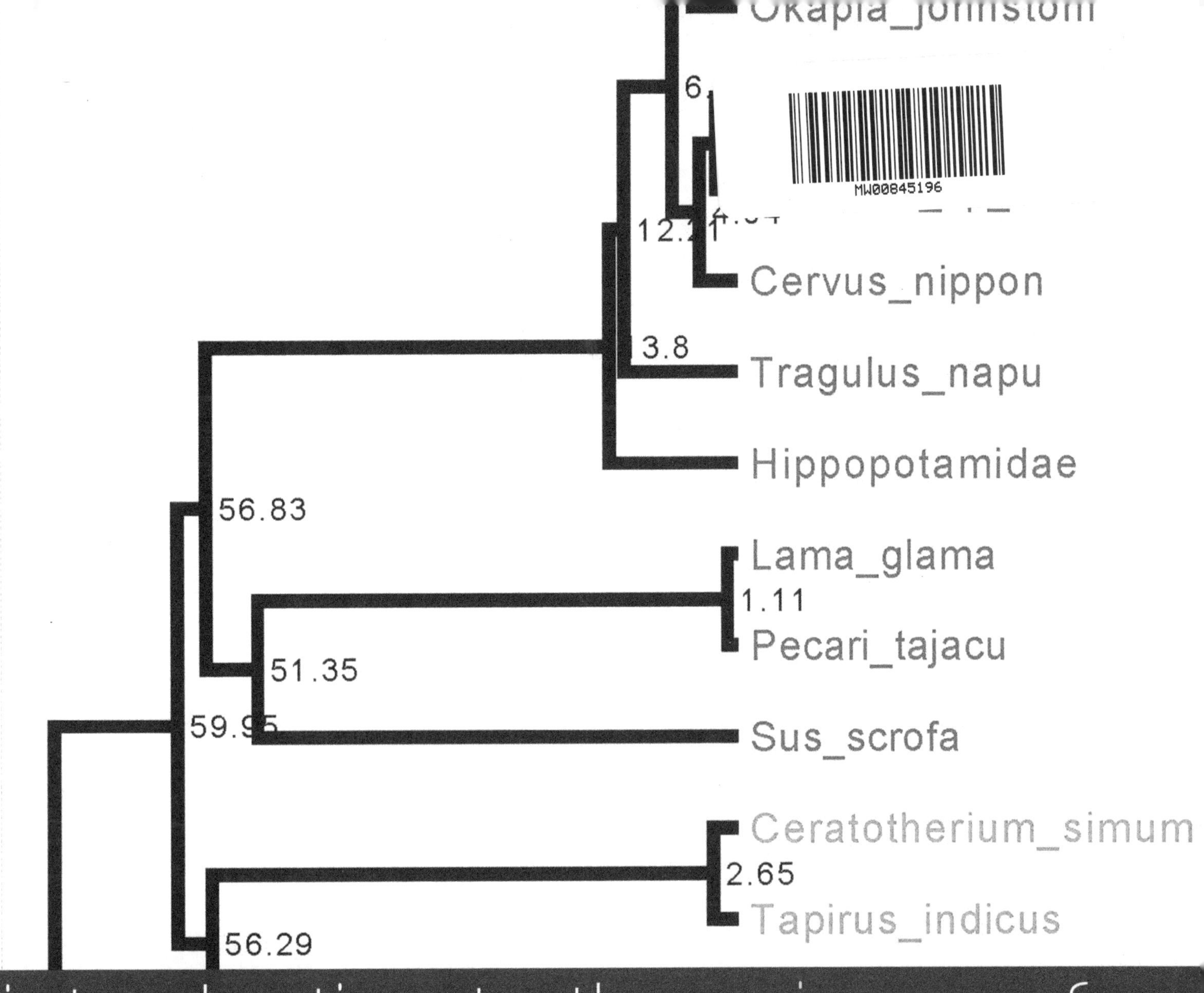

introduction to the science of

EVOLUTION

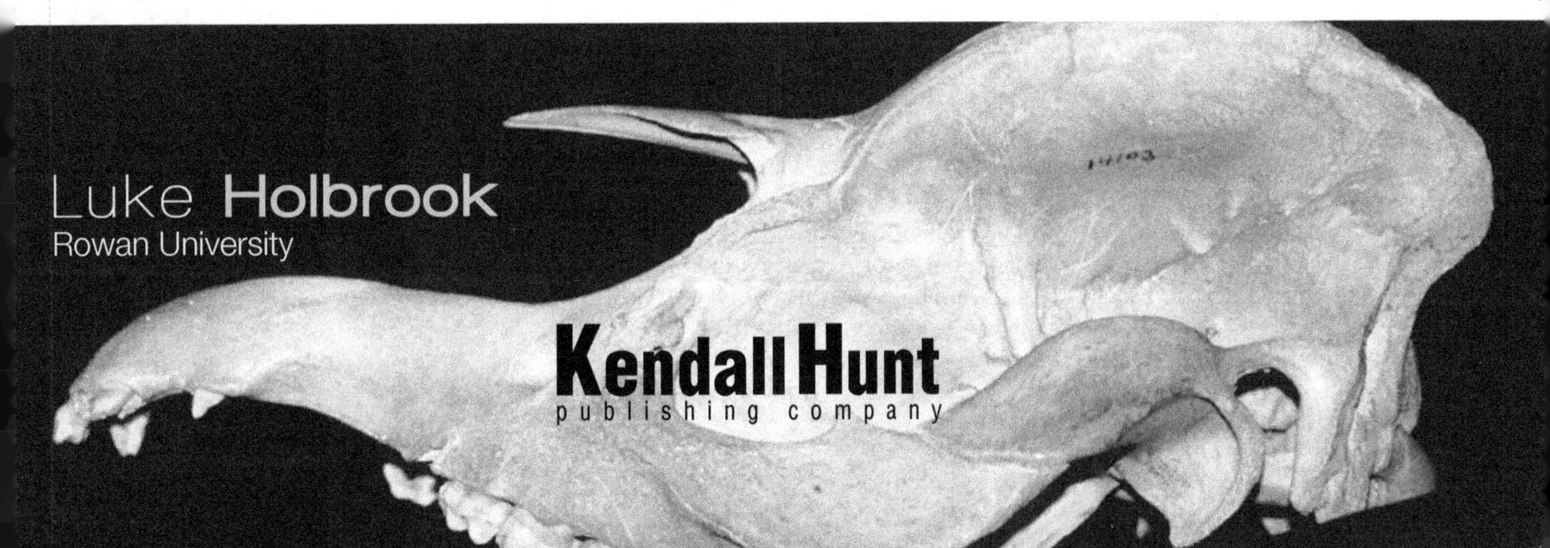

Luke Holbrook
Rowan University

Kendall Hunt
publishing company

Cover image © Shutterstock.com

www.kendallhunt.com
Send all inquiries to:
4050 Westmark Drive
Dubuque, IA 52004-1840

ISBN 978-1-5249-5565-6

Published in the United States of America

Contents

Introduction

Evolution is perhaps the central scientific principle of modern biology. It touches on every aspect of biology, from molecules, genes, and cells to communities of species and ecosystems. More than a century and a half of biological research has tested and validated evolutionary concepts, and evolutionary research has illuminated many things in biology that we otherwise could not have understood.

Evolution is also one of the most misunderstood concepts in biology, if not in all of science. The biggest obstacle to students learning evolutionary concepts is not necessarily that these concepts are abstract or arcane, as might be the case for, say, molecular biology or biochemistry. Rather, it is the fact that students often have a multitude of preconceived notions of what evolution is about and what it means. One such preconception is that evolutionary biology is fundamentally different from traditional laboratory science. Evolution, students often believe, deals with events occurring over long expanses of time in the distant past, which do not lend themselves to experimentation. As a result, students (and people in general) do not have a very good understanding of why evolutionary biology is a science, the same as molecular biology, biochemistry, or physics. The failure to appreciate the scientific merits of evolution makes it easier for individuals to accept arguments that evolution is not a science, or that it is "just a theory." Overturning this sort of thinking requires students to learn the scientific basis of evolution from the very beginning of their biology curriculum, in their first introductory biology course.

If you open up just about any introductory biology textbook, two of the things you will find in its earliest chapters are a discussion of the scientific method and a demonstration of evolution as a unifying principle—if not *the* unifying principle—of modern biology. It would be difficult to find a modern introductory biology textbook that understates the latter. The pages devoted to evolution in such texts offer not only an exposition of evolutionary concepts but also a demonstration of the evidence for why evolution is accepted by scientists as true.

These same introductory textbooks, because they are meant for initiates to science, often begin with a discussion of the scientific method. Increasingly, textbooks try to develop students' awareness of the practice and value of the scientific method by illustrating the experimental basis for why we accept some hypotheses and reject others. Cases are presented to show how experiments and other tests allow us to make predictions based on hypotheses. Matching predictions with experimental results allows us to evaluate the hypotheses.

Even though evolution is such an important concept in science, the foundational concepts of evolution are almost never presented in the context of the scientific method, at least not in the same way as the foundational concepts of, say, physics. Scientific theories, such as

Isaac Newton's laws of motion and atomic theory, don't stand on some transformation into "facts"; rather, they are still hypotheses in the sense that they are proposed answers to scientific questions, but they are hypotheses that have withstood a history of rigorous testing. In contrast, evolution is not typically presented as a hypothesis, or, more correctly, a set of hypotheses, that can be and has been tested. Evolution is demonstrated to be logical, and to be supported by particular evidence, but there is rarely a systematic demonstration of what evolution predicts and how that matches what we observe. In short, evolution does not get to show its scientific credentials.

The reason for this is perhaps a matter of perception. Evolution is seen as largely making statements about the past, and specifically the distant past. This seems beyond the pale of experimentation, so how can we present evolution as a testable concept? This book sets out to do just that.

The purpose of this book is to present evolution as a set of specific, far-reaching hypotheses; to illustrate how these hypotheses make specific predictions about the natural world; and to show how these predictions match what we find in nature, including in the results of experiments. As a result of surviving this testing, evolution has become a theory with extensive explanatory power. Chapter 1 sets out the scientific method, emphasizing the interplay of prediction and observation. The components of the scientific method are familiar to most students, but students often learn these components without learning how to use them. Instead, science comes across as something practiced only by professional scientists. One goal of this book is to have students engage in the process of science, and to understand that this process is not an activity restricted to people in lab coats with advanced degrees. It is something that we can apply to solving problems in our daily lives.

The next five chapters place evolution in the context of the scientific method. Chapter 2 introduces the notion of the testability of evolution, and divides the composite concept of "Darwinian evolution" into several component theories, which we treat as testable hypotheses. Chapters 3–6 each focus on a different component hypothesis and use a variety of cases to illustrate how their predictions match our observations.

The last three chapters are devoted to a brief introduction to the history of life on earth. Chapter 7 introduces the geological context for understanding the history of life. Chapter 8 discusses the science of understanding the origin of life. Finally, Chapter 9 is a whirlwind tour surveying biodiversity, with an emphasis on the groups of organisms that tell us about the ancestry of our species over the last four billion or so years.

How this book is organized for learning

The chapters in this book include much that is meant simply to be read and, hopefully, understood. But simply reading doesn't always translate into learning; students often need to *do* something before they truly begin to understand a concept. What they do—an activity—might be solving a problem, interpreting data, or even talking to other students. Thus, this book is meant not only to inform, but to get the reader to think, to discuss, and to actively engage in learning by making his or her own inquiries, proposing solutions, and testing them.

In this book, students will find activities associated with each important concept. Descriptions of activities are set off in boxes. The descriptions are fairly general, such that they can be implemented in a variety of classroom contexts. Of course, an instructor might provide additional exercises or might alter the activities provided here to suit the needs of the course.

Activities are presented in boxes in two forms. Some activities are labelled as "Brainstorm," because they are instructing the reader to simply spend some time thinking about something, and perhaps discussing the topic with classmates. Brainstorms are meant to get the reader to come up with some of the information and concepts that are discussed in the chapter, before proceeding to the next page for an answer. Other activities are labelled as "Activity" and have specific instructions for coming up with solutions to particular problems. Ultimately, the answers to these problems are discussed later in the chapter itself, but it is expected that the reader will make the effort to try to solve the problem on his or her own, rather than simply reading ahead. While the latter might be tempting, only through engaging in the learning process does the student truly come to understand how solutions arise. Furthermore, these activities are meant to follow the methodology of science, so a reader that engages in the activities enhances his or her training as a scientist.

There is one other type of "box" used in this text, beginning with the title "A Closer Look." These boxes review information that is important for understanding the concepts discussed in this text, but that is not central to the theme of this book. Such boxes include discussions of atomic theory and radiometric dating, Mendelian genetics, and the structure and function of DNA, to name a few. Because this is a book designed for students early in their college training for biology, I don't assume that the intended reader already has a background in some of these fundamental biological and scientific concepts. Thus, the boxes provide a resource for students previously unfamiliar with these concepts, or for students who have some familiarity but who could use an opportunity to review this information.

Acknowledgements

This book grew out of teaching a class for first-semester freshmen Biology majors at Rowan University called Introduction to Evolution and Scientific Inquiry. While I have been the lone instructor for my own sections of this course, there are many other instructors for other sections, and I have had the good fortune to work with a number of colleagues in the Department of Biological Sciences who developed course materials and approaches to topics that have influenced my own teaching and this book. In particular, I am indebted to Mike Grove, Matt Travis, Matthew Bealor, Amy Combs, and Frank Varriale for discussions about this course and for providing cases that illuminate many of the concepts presented here. I also am grateful to Courtney Richmond for reviewing an early draft of this book and trying to set me straight on statistical topics. Courtney, Mike, and Alison Krufka have also been integral to my own development as an instructor and helped me discover the value of active, student-centered, inquiry-based learning.

Part I

Applying the Scientific Method to Evolution

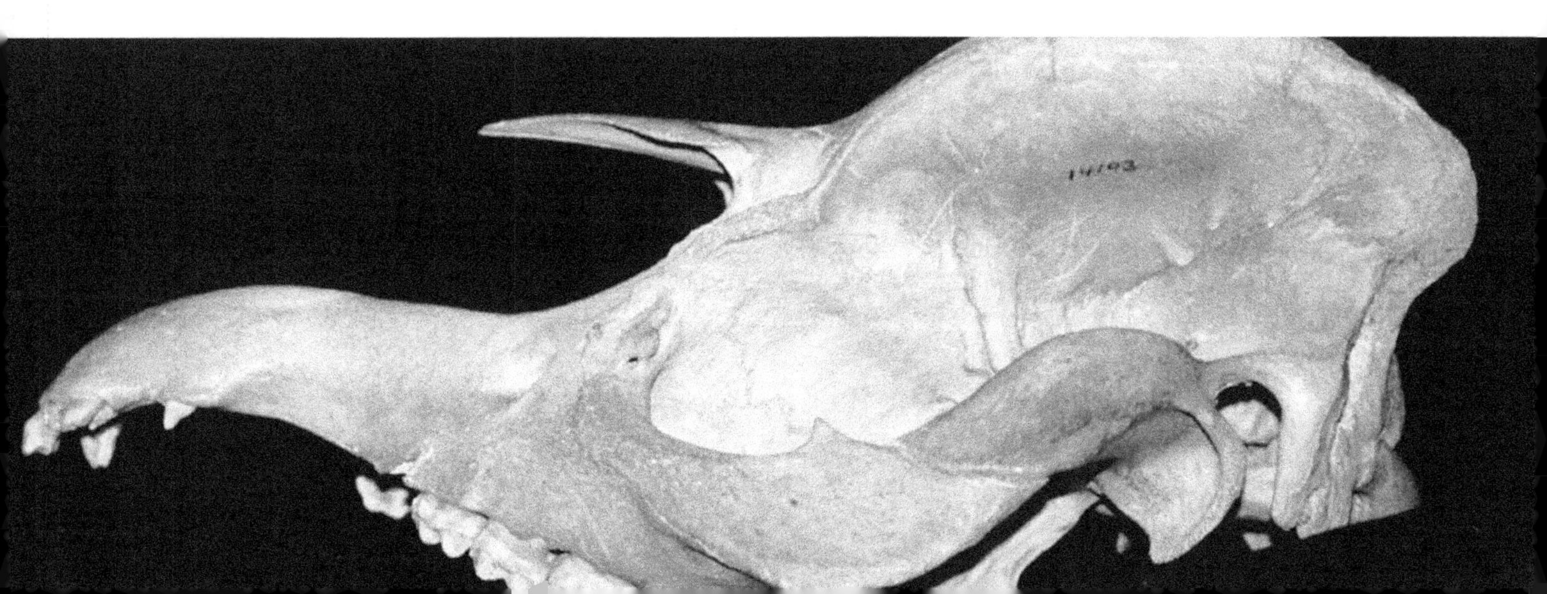

CHAPTER 1 The Nature and Process of Science

This chapter outlines the scientific method in a way that sets up how we can apply this method to evolutionary concepts. Our understanding of the natural world consists of the limited number of things we observe and our conjectures as to what else fills in the gaps. New scientific observations are being made all the time, many of them as a result of laboratory experiments and field studies. The conjectures are our hypotheses of the causes behind the patterns we perceive in the natural world, as well as our predictions of what else we will observe when we make further investigations. Good scientific hypotheses make predictions about what we should find in the world, and we can test hypotheses by seeing if their predictions match what we observe, whether these observations are the results of a laboratory experiment or discoveries in the fossil record.

1.1 Questions, patterns, and the natural world

We usually begin discussion of the scientific method with making an **observation** that stimulates the formation of a **question**. This is actually one of the biggest stumbling blocks for students, as they feel unable to identify questions to investigate. In fact, our daily lives are filled with questions to which we can apply scientific thinking. Issues as mundane as a car that won't start lead you to wonder why, then to hypothesize what might be wrong, and subsequently to think of other observations you can make to test that hypothesis.

Questions can come from things that directly affect our lives, but most scientists ask questions about broader patterns that we perceive in the natural world. Why do plants grow? Why are there seasons? As we think more about natural phenomena, new questions present themselves. In many cases, more specific questions may lead a scientist to ask more general questions. Why do polar bears live in the Arctic but not in Antarctica? This might require answering the broader question of why the poles don't have the same species living in them when they are such similar environments.

Brainstorm

Working in groups, come up with as many questions as you can regarding natural phenomena. (Remember: natural phenomena include everything that we can observe, so even things you might not think of as "natural"—things made by or concerning humans, for instance—are actually part of the natural world.)

1.2 Hypotheses

Questions naturally lead to wondering about answers, and our attempted answers are **hypotheses**. Good hypotheses have certain qualities. They are usually about **causation**; in other words, they propose causes or reasons for why a phenomenon exists. Returning to the example of why a car won't start, a possible reason that this is true—the car is out of gas or the battery is dead—would be a hypothesis.

When we think of causes, we might be interested in whether two things that we can measure are related (or **correlated**, as we'll discuss in a later chapter), where, say, when X is high, Y is also high, and when X is low, Y is also low. But is this relationship causal? Does X determine Y, or vice versa (or neither)? For instance, we might find our car that won't start also has a radio that doesn't function. That doesn't necessarily mean that the radio is the reason the car won't start. Good hypotheses typically try to get at these causes, rather than just vague statements of relationship.

Good hypotheses are **testable** and **falsifiable**; they "stick their necks out," so to speak, and can be subjected to testing that can result in rejection. Why do we value the possibility of rejection over the possibility of proof? This is because it is logically possible to disprove a hypothesis but not to prove it, a concept that comes from the work of the philosopher of science, Karl Popper. For instance, take the hypothesis "X causes Y." I could then run an experiment where I measure X and Y and predict, based on the hypothesis, that I should get high values for Y when X is high, and low values for Y when X is low. If I get no such relationship between X and Y, I can reject my hypothesis. If I get the predicted relationship, it is consistent with my hypothesis, but have I truly ruled out the possibility that some third phenomenon (call it Z) is driving the values of X and Y?

Again, we can illustrate this with our example of the car that won't start. If my hypothesis is that the battery is dead, I can make predictions about other things that should be true if the battery is dead, such as that the radio should not work. If the radio doesn't work, that supports my hypothesis, but it doesn't rule out the possibility that the radio is not working for some other reason. On the other hand, if the radio does work, then I positively know that the battery can't be dead. A working radio falsifies the dead battery hypothesis.

We can falsify hypotheses, but if we can't actually prove a hypothesis, what, then, allows us to elevate a hypothesis to being an accepted explanation for something? It has to consistently defy rejection, and it has to do so better than competing hypotheses.

Ultimately, we want to compare hypotheses to see which one is better supported by our data. Even if we haven't thought of two separate possible answers to our question, we can still compare our one hypothesis to the **null hypothesis**. The null hypothesis is essentially the negation of the hypothesis we are considering (which we will call the **alternate hypothesis**). That does not mean that the null hypothesis is the opposite of the alternate hypothesis; rather, if we are hypothesizing that X causes Y, our null hypothesis would be that X does not cause Y, not that Y causes X. For instance, if we were considering the alternate hypothesis that hot dogs cause earthquakes, our null hypothesis would be that hot dogs do not cause earthquakes, or that hot dogs have nothing to do with earthquakes. It would not be that earthquakes cause hot dogs!

Since we will be referring to hypotheses very often, we will use the letter H for "hypothesis," with a subscript to identify a specific hypothesis. Often, we'll simply use H_A to represent the alternate hypothesis and H_0 (subscript "zero") for the null hypothesis.

Activity

Come up with hypotheses to answer some of the questions you developed in the brainstorming in the previous section, or consider sample hypotheses provided by your instructor. Are they testable? Are they falsifiable?

Come up with null hypotheses for your hypotheses, or for hypotheses provided by your instructor.

1.3 Predictions

Presumably our hypothesis does a good job of explaining whatever we observed that caused us to ask a question; in other words, if it were true, our hypothesis would fully answer our question. We can make **predictions** based on that hypothesis regarding what else we might observe. Ideally, if those predictions don't match what we observe, we can reject our hypothesis. In the car example, we made a prediction about the functioning of the radio based on the hypothesis that a dead battery was preventing the car from starting; we also could have made a prediction that the radio would work based on the null hypothesis that the battery had nothing to do with why the car wouldn't start. In reality, data rarely fit a hypothesis perfectly, due to various sources of error, so what we do instead is compare how competing hypotheses fit our data and reject the hypothesis that has the worst fit. If we're only considering a single hypothesis, we can compare its predictions to those of the null hypothesis. When we do an experiment, we essentially attempt to manipulate conditions such that results predicted by the hypotheses being compared will be different.

For instance, consider the hypothesis that plants need sunlight to grow. The null hypothesis would be that sunlight has no effect on plant growth. We could then make predictions about how plants will grow with and without sunlight. According to our alternate hypothesis, plants would grow much better in sunlight than they would in darkness. Our null hypothesis would predict that plants will grow about the same regardless of whether they have sunlight or not. Thus, the alternate and null hypotheses predict that we will observe different things depending on which hypothesis is true.

Note that when we make predictions, we often need to specify exactly what the observations would be and how we would measure them. For instance, in our example above, we made predictions about growth based on each hypothesis. But what do we mean by "growth"? How would we know if one plant has "grown" more than another, and how could we represent this to another scientist? To solve this, we could take a measurement that we think describes growth. We could measure the height of a plant, or count its leaves, or weigh it. But "growth" refers to how much the plant has changed during the experiment, so we could take the measurement at the beginning and the end of the experiment and determine the difference between the two; the value of the difference would be our measurement of growth. What we have constructed is a measurement to use as our **response variable**; in other words, this is the variable that we expect to have different values (i.e., to "respond" differently) in the experiment depending on whether the alternate or null hypothesis is true.

The response variable can also be called the **dependent variable**, because, at least according to the alternate hypothesis, its value should be dependent on another variable. In the case of the plant experiment, our dependent variable of growth is determined by the extent to which the plant is exposed to sunlight. The amount of sunlight would be our **independent variable**. You will often see graphs depicting the relationship between an independent and a dependent variable; the convention is for the *x* axis to represent the independent variable and the *y* axis to represent the dependent variable.

Activity

For each of the examples of hypotheses and related experiments provided below, do the following:

A) *Determine the null hypothesis.*

B) *Make predictions regarding the results of the experiment based on the alternate hypothesis and on the null hypothesis. What would the response variable(s) be for each experiment?*

C) *Make graphs representing your predicted results. What would go on the axes of your graphs? Which of the variables are the dependent and independent variables? You might not be able to specify the exact values predicted for the data for each hypothesis, but you should be able to depict how they would compare on the graph.*

Sample hypotheses with brief descriptions of experiments

1. *Acetylcholine stimulates muscle contraction.*

   Cultures of muscle fibers are prepared. A solution of acetylcholine is applied to some of the cultures at different concentrations. Other cultures receive only the solvent without acetylcholine, and still others have nothing applied to them.

2. *Water moves from the roots of a plant upwards because of transpiration, where evaporation of water from leaves draws water up through the xylem.*

   Plants are placed in a medium where their roots are submerged in water with a dye. Some plants have their stomata, tiny openings on the undersides of the leaves that allow for gas exchange, painted shut with nail polish. Other plants have some leaves painted, others not painted. Still other plants are not painted at all.

3. *Amphibian metamorphosis is regulated by the thyroid hormone thyroxin.*

   Tadpoles are kept in a number of tanks. Some tanks have the thyroid hormone thyroxin added at different concentrations. Other tanks have no thyroxin added.

1.4 Tests and their design

We have discussed how hypotheses can lead to predictions, and comparing predictions to observations is ultimately how we test hypotheses. If the battery is dead, we predict that the radio will not work, and if the battery is not dead the radio will work. Those predictions provided a simple test of our hypothesis. Most tests of scientific hypotheses are more complex. Hypotheses make many predictions, but which predictions will be most helpful for a test? What other things could produce the observations that we predict for our hypothesis, and how can we discriminate between causes related to our hypothesis and other kinds of causes? What do we measure and how do we measure it in order to know whether or not our predictions have been met? Even if we can measure something, how do we know if it has changed in the way we predicted or not? The answers to these questions are all part of designing a good test of a hypothesis.

We typically call these well-designed tests **experiments**, but note that there are different kinds of experiments. The word "experiment" often conjures up images of antiseptic laboratories, glassware with graduated measurements, and people in white coats. Certainly, many experiments are performed in labs, because labs provide environments amenable to good experiments. But experiments can—and sometimes must—occur outside of the lab.

Good experiments have a number of qualities. First and foremost, the results of the experiment will be different depending on whether one hypothesis or another is true. In other words, in a good experiment, the null hypothesis will predict different results than what is predicted by the alternate hypothesis. If the hypotheses make the same predictions, then the test will not allow us to determine whether one hypothesis fits the data better than the other. We can then ask whether the actual results match the predictions of one hypothesis or the other, or perhaps even of neither.

Good tests also provide a way to minimize the effects of factors other than the one about which we are making predictions. This is the function of a **controlled experiment**, where we control all of the relevant variables in an experiment, whether they pertain to the cause we are testing or to other factors. Controlled experiments also typically include **control treatments**. Control treatments differ from **experimental treatments** in that they have not been manipulated in terms of the variable that is being tested. For instance, in our car example, we could control for other causes of a radio not working. We could start the engine and turn on the radio of another car, which would allow us to know what we should hear on the radio if it is working. We could replace the car's battery with a new one (or one known to be functioning) and see if the radio is working.

As another example, let's say you wanted to test whether a particular drug has a particular effect on mice. You could administer the drug to mice in your experimental treatments but not to those in your control treatments, and all mice in all treatments would otherwise be as similar as possible. This way, if the mice in the experimental treatments exhibit effects that differ from those in the control treatments, we can have the greatest confidence that those differences are due to the effect of the drug.

In essence, the control treatments give us a point of comparison for our experimental treatments. In the plant experiment, our experimental treatments were the plants we placed in the dark. We predict that these plants would not grow as well as our control plants, which were the ones in the sunlight. Without the control plants, we would have no way of evaluating whether what happened to the plants in the dark was really affected by the lack of sunlight.

Activity

Using the three hypotheses and experiments in the previous activity, do the following:

Identify the control treatment(s).

Use the results given in the tables below to determine whether the data were consistent with the predictions of the alternate or null hypotheses.

1. *Acetylcholine experiment*

Type of treatment	Percentage contracting
Acetylcholine added	100
Solvent only	0
No acetylcholine or solvent	0

2. *Water movement in plants*

Type of treatment	Movement of water (as percentage of plant height)
All stomata covered	0
Half of stomata covered	50
No stomata covered	100

3. *Amphibian metamorphosis*

Type of treatment	Percentage metamorphosing
Thyroxin added	85
No thyroxin added	2

There are additional qualities of a good experiment. Good experiments include more than one subject, to ensure that the results are not biased by the peculiarities of a single individual; the number of subjects is often what we refer to as the **sample size**. Most experiments strive to have a large enough sample size to make comparisons meaningful and to enhance the utility of any statistics that are applied to the data.

Good experiments also account for other **confounding variables** that could be affecting the results, particularly those pertaining to **environmental variation** and **individual variation**. Environmental variation includes a number of conditions pertaining to the environment in which the experiment is run. These would include things like temperature, air pressure, and lighting conditions, but also things like the nutrition and habitats of living subjects, the timing of treatments, and other conditions that are external to the subjects themselves. Ideally, these environmental conditions will be the same for all subjects in the experiment; otherwise, we might not be able to eliminate the possibility that a difference in the results obtained from different subjects is due to differences in their environments. This is why experiments are often performed in labs, where the environment is carefully controlled and maintained to be as identical as possible for all treatments.

Individual subjects are not identical, so how do we determine whether a difference between the results from two subjects is not due to the fact that they are simply different individuals? Good experiments therefore need to control for individual variation. For instance, many experiments use model organisms, such as specially bred strains of mice or bacteria, where all individuals are as close to being genetically identical as they can be. Even if your mice are genetically identical, they might still be different due to the different lives that they have led. Thus, you would also want to make sure that your mice were of similar size, age, and health.

In many cases, you cannot simply select subjects that are nearly identical. This is especially true for clinical studies of human health. How does one control for individual variation then? In such cases, a researcher can design an experiment to minimize **bias**, or some kind of nonrandom variation in the treatments. For instance, if your control mice were all males, but your experimental mice were all females, that would be a very biased distribution of sexes in your treatments, and you would not be able to exclude the possibility that the differences in the results between your controls and experimentals were due to sex differences. In experiments like clinical trials, bias is avoided by **randomization**; subjects are assigned randomly to a treatment group, and with large enough sample sizes each treatment should have a set of individuals who, while not identical to each other, vary in roughly the same way from treatment group to treatment group.

Finally, good tests are or can be repeated, by the original researcher or by those in another lab. The terms **repetition** and **replication** are often used loosely for either kind of repeating an experiment, but typically repetition refers to essentially having multiple instances or runs of an experiment when it is carried out, whereas replication typically refers to someone else carrying out the same experiment to see if they get similar results. Note that, in this case, repetition might mean running an experiment several times in sequence, or having multiple control and treatment groups. In many cases, simply using multiple subjects is a form of repetition.

Activity

A) *For the three experiments in the previous activities, identify the additional information you would need if you were going to run these experiments yourself.*

B) *What would you need to do in each of the experiments to execute them and to control for confounding variables?*

C) *How many individuals would you need, and how would you take into account the effects of differences between individuals?*

A Closer Look: Descriptive Statistics and Calculating Means and Standard Deviations

An important part of experimental design is determining what you will measure and how you will measure it. After you make your measurements, how will you interpret them? For instance, in the plant experiment, when we measure the heights of our plants, it is very likely that they will not all be the same height, but how different should they be for us to say whether the prediction of a hypothesis is matched or not? If we use many plants in each treatment, we will need a way to describe the growth of each treatment as a whole. Descriptive statistics help us to do that. Two of the most basic descriptive statistics are the mean and the standard deviation. The **mean** is the average value of a measurement (say, of a particular trait) for all of the sampled individuals. While the mean is helpful, it doesn't provide a very complete picture of variation in this trait in the population. In other words, we might glean something from the fact that the mean is large or small, but is that mean derived from a sample that varies very little from the mean or that has a wide range of variation in this trait? A useful statistic for understanding the variation of a trait among sampled individuals around its mean is the **standard deviation**. The standard deviation is essentially the average amount by which individuals vary from the mean. If the range of variation is wide, then the standard deviation will be large relative to the mean; if there is little variation in the sample, then the standard deviation will be small relative to the mean.

The formulas below are for calculating means and standard deviations.

Mean: $\overline{X} = \frac{\sum x_i}{n}$ Standard deviation: $\sqrt{\frac{\sum (x_i - \overline{X})^2}{n-1}}$

The mean is fairly straightforward and probably familiar to most students, even if its representation in the above formula is not. The mean takes the sum of all the individual values for a given variable (in this case, x, with each individual's value for x given as x_i); that sum is Σx_i in the formula. We then divide that sum by the number of individuals in the sample, or n.

The formula for standard deviation shows some similarities to the formula for the mean: there is a sum being made that involves x_i, and we are dividing by a number close to the sample size ($n-1$). The main differences are that (a) we are subtracting the mean from x_i, (b) we're squaring that difference, and (c) we're taking the square root of the whole calculation. The first difference should make sense: we are interested in how individuals vary from the mean, on average, so we can start by calculating the difference between each individual's value for x and the mean. But why square this difference? The reason is that individuals vary both by being less than the mean and by being greater than the mean. Thus, if we just take differences from the mean, we will end up with some positive numbers and some negative numbers, and the sum of those will be zero. By squaring the difference, we eliminate the sign, and all squared differences will be positive. Thus, the resulting sum will be positive. By taking the square root, we are essentially "undoing" that squaring and ending up with a number that reflects how the population tends to vary from the mean.

The standard deviation is useful for describing how much a sample varies. Here is a simple example to illustrate this.

	Tree heights in area 1 (m)	Tree heights in area 2 (m)	Tree heights in area 3 (m)	Tree heights in area 4 (m)
	5	3	1	1
	5	3	2	1
	5	3	3	1
	5	5	4	5
	5	5	5	5
	5	5	6	5
	5	7	7	9
	5	7	8	9
	5	7	9	9
Average height (m)	5	5	5	5

Note that all four populations have the same mean. But is the variation in the populations the same? We can assess that using the standard deviation.

	Tree heights in area 1 (m)	Tree heights in area 2 (m)	Tree heights in area 3 (m)	Tree heights in area 4 (m)
	5	3	1	1
	5	3	2	1
	5	3	3	1
	5	5	4	5
	5	5	5	5
	5	5	6	5
	5	7	7	9
	5	7	8	9
	5	7	9	9
Average height (m)	**5**	**5**	**5**	**5**
Standard deviation (m)	***0***	***1.63***	***2.58***	***3.27***

Note the differences in standard deviation. The population where all individuals are the same has a standard deviation of zero, whereas those that have individuals that are very short and others that are very tall have larger standard deviations. Thus, the populations with more variation have larger standard deviations.

1.5 Extending tests beyond the lab

Typically, we think of tests as **manipulative** experiments; in other words, the scientist manipulates the conditions of the experiment, creating a situation that does not exist outside of the lab, and which allows for the greatest control of confounding variables. Indeed, the controlled experiment is the standard for testing in science. In some cases, controlled experiments are not feasible, but we may still be able to test hypotheses through **observational experiments**, using comparative methods and "natural experiments." For instance, if we wanted to test the hypothesis that the number of species on an island is determined by the size (area) of the island, we could compare the number of species on lots of islands of different sizes. Although we can't create a control treatment for such a study, we can still try to control for confounding variables by how we sample islands. For instance, we can restrict our sample to islands within a certain range of latitudes, in order to control for the effect of latitude.

As another example, consider the three-spined stickleback (*Gasterosteus aculeatus*; Figure 1.1). These fish are found in marine and freshwater environments in the northern part of the Northern Hemisphere; marine populations have managed to establish themselves in inland lakes and streams numerous times. In the marine populations, the fish are "armored" with bony plates and spines along their flanks and undersides. Some freshwater populations retain this armor, but in others this armor is greatly reduced or even absent. One hypothesis is that the armor, which costs the fish considerable energy to make, is an adaptation for protection against predators. Based on this hypothesis, we would predict that sticklebacks would be well-armored when populations are in places with many predators. If we looked at different lakes with sticklebacks, we would predict that we'd find armored fish in the lakes with predators and unarmored fish in those that lack predators. We would not be manipulating the lakes or the fish populations, but instead we would rely on the natural occurrence of lakes that have sticklebacks but that differ in the presence or absence of predators.

Jack perks/Shutterstock.com

Figure 1.1. A three-spined stickleback (*Gasterosteus aculeatus*).

Observational experiments obviously lack some of the opportunities for controlling and manipulating conditions that manipulative experiments enjoy, but that is not to say that observational experiments cannot be as powerful, and they are often necessary. Nor are observational experiments only typical of studies involving observations of plants and animals in the field. Much of what we know about astronomy comes from careful observation of the night sky. Many studies important for our understanding of human health are actually observational studies. A prominent example involves the investigation of the alleged relationship between the measles-mumps-rubella (MMR) vaccine and autism in children. Several observational studies in Finland, Denmark, and the United States examined vaccination and hospital records looking for correlations between autism and vaccination, including the timing of both. In some cases, the researchers compared the incidence of autism in children who had been vaccinated with that of children who had not been vaccinated; the latter group acted as a control group. No significant difference was found in the incidence of autism in children who had and who had not been vaccinated, and there was no relationship between when children were vaccinated and the development of autism. Despite the fact that this experiment could not be carried out in a laboratory, it still provides conclusive evidence regarding the lack of a causal relationship between vaccines and autism.

1.6 Where do we go from here?

An elegant test might allow us to reject competitors and settle on a particular hypothesis. Regardless, while a hypothesis can be falsified, it can be supported but can't be deductively proved to be absolutely true, so there may still be room for more testing. Other scientists may repeat a test to see if they can corroborate the results. More often, answering one question leads to other questions, and the process begins again.

CHAPTER 2 The Theories, Predictions, and Tests of Evolution

From its birth with the publication of *Origin of Species*, Darwinian evolutionary theory has been challenged, not only by those who felt it threatened their personal beliefs, but also by the investigations of Darwin's contemporary scientists. This chapter introduces the notion of the testability of evolution with a historical example. We also see how evolution can be viewed as a composite of several theories, each of which makes its own predictions and allows for additional tests of Darwin's ideas.

2.1 Testing evolution: an illustration using the age of the earth

Perhaps, the best example of the testability of evolution comes from nineteenth-century efforts to estimate the age of the earth, most famously by William Thomson, also known as Lord Kelvin (Figure 2.1). Thomson estimated the age of the earth to be around 20 million years, based on his knowledge of thermodynamics. This was too young a planet for Darwin's ideas to explain the diversity of life, and it was perhaps the most legitimate scientific threat to Darwinian evolution during his lifetime (though Darwin thought these estimates were likely to be wrong). As it turned out, Thomson's estimates were incorrect, because he (like everyone before 1896) was unaware of radioactive decay, which makes the earth warmer than if it was simply a formerly molten rock cooling in space. (He also based some of his estimates on calculations of the age of the sun, which he underestimated because nuclear fusion that powers the sun was not discovered until the twentieth century.) Modern estimates of the age of the earth are about 4.6 billion years, which is consistent with the predictions of Darwinian evolution.

Thomson's attempt to calculate the age of the earth demonstrates something that is often not appreciated: that evolution is testable. In proposing his evolutionary theory, Darwin had predicted that the earth was actually hundreds of millions of years old. Investigators like Thomson (and including Darwin's own son, a physicist whose calculations produced results roughly similar to those of Thomson) were testing this prediction by trying to determine the age of the earth. These early calculations were not accepted as conclusive, so their contradiction of Darwin's predictions were not considered as falsification of Darwin's ideas. Only in the twentieth century, when radiometric dating allowed for more accurate calculation of the age of the earth, did Darwin's theory face a critical test, and this time the results were consistent with Darwin's prediction.

Figure 2.1. Lord Kelvin, also known as William Thomson.

A closer look: Radioactivity and determining the age of the earth

Even in Darwin's time, scientists knew that not all rocks were of the same age, and the **relative ages** of rocks could often be established. It was clear that some rock units had been formed on top of other layers of rock, indicating that the lower rocks had to be older than the upper rocks. But how much older? How can we establish an **absolute age**—in other words, an actual (if approximate) number of years old—for a rock?

Atoms and their structure

The theory that matter was composed of tiny units called **atoms** existed in Darwin's time, but atomic theory was still in its infancy until the discovery of subatomic particles at the end of the nineteenth century. These discoveries established that atoms were not indivisible but in fact consisted of some combination of three types of subatomic particles, illustrated in Figure 2.2: light, negatively charged **electrons**; heavy, positively charged **protons**; and heavy **neutrons** with no charge. The number of protons in an atom determines what **element** it is, and each element is identified by an **atomic number** equal to its proton number. Hydrogen, for instance, has one proton, whereas carbon has six, and their atomic numbers are 1 and 6, respectively.

Atoms of the same element can vary in the number of neutrons; these different varieties of an element are called **isotopes**, which differ from each other in **atomic mass**, or the number of protons plus the number of neutrons. For instance, carbon typically has six protons and six neutrons, for an atomic mass of 12 (and thus this isotope is called carbon-12). Carbon-13 and carbon-14 are much rarer isotopes of carbon that have seven and eight neutrons, respectively. The atomic mass of carbon is usually reported as something like 12.011, which is the average of the atomic mass of all carbon atoms, taking into account the abundance of different isotopes. Figure 2.2 illustrates some examples of isotopes for different elements.

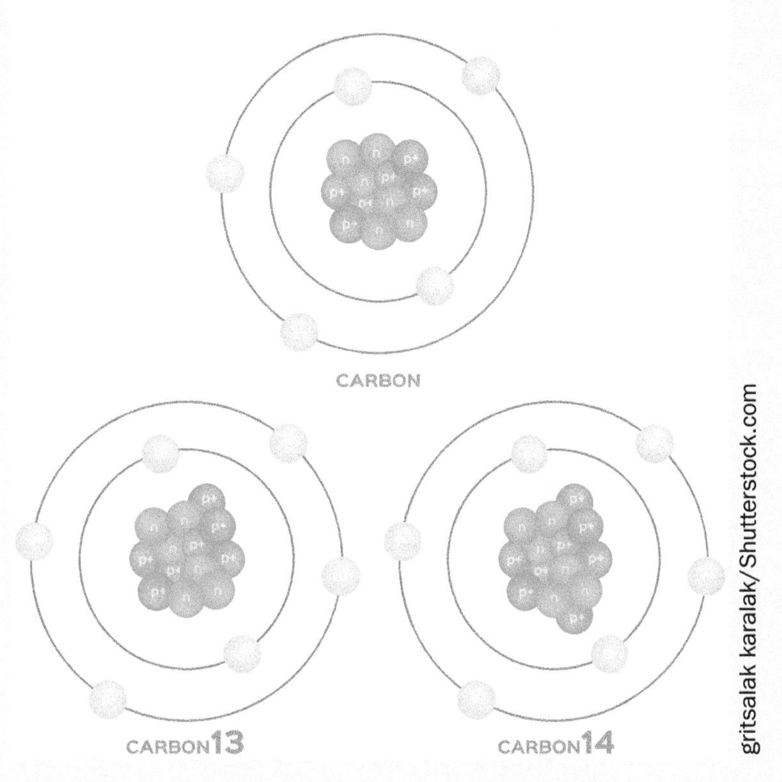

Figure 2.2. Atoms, specifically isotopes of carbon, showing protons (pink), neutrons (yellow), and electrons (blue). Note that each of these atoms has six protons, identifying each as an atom of carbon, but different numbers of neutrons, which makes them distinct isotopes.

Radioactive isotopes

Not all isotopes are stable. In other words, there are isotopes of some elements that will exist only temporarily and eventually transform in some way. These are **radioactive isotopes**, also called **radionuclides**. The transformation, or **decay**, of a radioactive isotope typically results in some change in the nucleus. The nucleus could lose neutrons or protons, or a proton might be transformed into a neutron, or vice versa. If there is no change in the number of protons, the radioactive isotope (termed here the **parent radionuclide**) produces a **daughter nuclide** that is another isotope of the same element as the parent. If there is a change in the number of protons, then the daughter nuclide will be a different element from the parent. Carbon-14, for instance, decays into nitrogen-14. Note that the daughter nuclide might itself be a radionuclide and undergo radioactive decay on its own. Radioactive decay also results in the release of energy in the form of some type of radioactivity. The energy released from radioactive decay deep within the earth is what keeps it so hot inside, and Thomson's ignorance of this caused him to consider the earth to be much younger than we now know it to be.

Radiometric dating

Radioactivity and radioactive decay are physical processes that occur in a very predictable fashion. Half of a given amount of a radionuclide will decay into its daughter nuclide in a period of time specific to that radionuclide; this time is called the **half-life** of the radionuclide. The half-life of carbon-14 is calculated as about 5730 years. The half-life of uranium-238 is approximately 4.47 billion years.

Thus, if we start with a sample that consists entirely of a particular radionuclide, after a period of time equal to the half-life of the parent radionuclide, half of the sample will consist of the parent radionuclide and the other half will be the daughter nuclide; if I start with one mole of carbon-14, after 5730 years I will have half a mole of carbon-14 and half a mole of its daughter nuclide, nitrogen-14. After two half-lives, the radionuclide will be reduced to one-quarter of its original amount, and three-quarters of the sample will be the daughter nuclide. Three half-lives will result in one-eighth parent radionuclide and seven-eighths daughter nuclide, and so on. Figure 2.3 illustrates this process and its results.

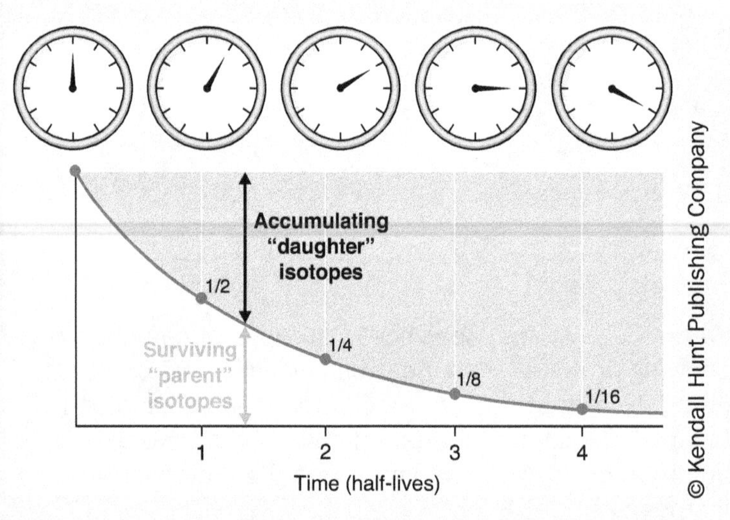

Figure 2.3. Radiometric dating. The graph illustrates how the relative amounts of the parent radionuclide and the daughter nuclide will change over time, measured in half-lives of the parent radionuclide.

If we examine a sample of rock that contains radionuclides, we can compare the relative amounts of the parent radionuclide and its daughter nuclides and use that information to estimate the age of the rock based on the parent radionuclide's half-life. The process of analyzing the amounts of parent radionuclide and daughter nuclide, which might be naturally present in only small amounts in the rock, is often conducted using a **mass spectrometer**, which can identify the different elements and isotopes present in the rock sample and quantify their abundances. Thus, based on how many half-lives it takes to get the calculated amounts of parent and daughter nuclides, we can calculate the approximate age of the rock sample in years. This method is called **radiometric dating**. The actual radionuclide chosen for analysis for radiometric dating depends on the hypothesized age of the sample in question. Carbon-14 is commonly used for archaeology, as it can provide reliable dates from carbon-containing items (including things like wooden objects and bones) as old as 45,000 years. However, carbon-14 is not very useful for anything older than about 45,000 years, because the amount of parent radionuclide is so small that the accuracy of the instrument is not sufficient to give a reliable date. Also, at some point no more carbon-14 is present, so all samples above a certain age would look the same, whether they are 500,000 years old or 500 million years old.

For determining ages of rocks formed when the earth was very young, researchers use radionuclides with very large half-lives, like uranium-238 and potassium-40. Uranium-238, as mentioned before, has a half-life of about 4.47 billion years, whereas that of potassium-40 is about 1.25 billion years. While these radionuclides are excellent for measuring the age of rocks formed early in earth history, they would be poor choices for archeological samples, which typically are from the last 100,000 years of earth history, and usually as recent as the last 10,000 years. Even a rock that formed 100,000 years ago would have had so little of its uranium-238 and potassium-40 decay that the amount of daughter nuclide would be too small for the mass spectrometer to accurately measure.

Estimating earth's age using radiometric dating has involved sampling a number of sources. The oldest rocks on earth have been dated to over 4.4 billion years old, but, because rocks are frequently eroded and transformed over time, that does not mean that these rocks are as old as the earth. Instead, researchers have examined objects that are thought to have had their origin at around the same time as the earth, such as moon rocks and meteorites, and from these sources we estimate the age of the earth to be about 4.54 billion years old.

Like any other scientific hypothesis, evolution makes predictions about the natural world, and we can judge it by how well its predictions match our observations. This is something that may come as a surprise to many people, because evolution is seen as an idea that governs phenomena that happened long ago in the past and that can't be replicated in the laboratory. In fact, as a far-reaching theory with implications for all aspects of biology, evolution makes a wide range of predictions that allows for a wide range of tests.

2.2 The five theories model of Darwinian evolution

It is easier to understand the kinds of predictions made by evolution if we think of Darwinian evolution as a collection of related, individual hypotheses or theories. This is exactly what Ernst Mayr did in his book *One Long Argument*, and we will adopt his model of looking at Darwinian evolution as the following five theories: evolution per se, natural selection, multiplication of species, common ancestry, and gradualism.

The five theories, illustrated in Table 2.1, can be briefly defined as follows. **Evolution per se** (or evolution as such) refers to the basic notion that lineages of organisms change over time; that is, that they evolve. **Natural selection** is Darwin's mechanism for evolution, where traits that increase an individual's chance for survival (and for reproduction) are favored and increase in their frequency in the population over time. **Multiplication of species** is the notion that one species can give rise to two or more descendant species, thus allowing a lineage to diversify over time, as well as for today's biodiversity to have arisen from an ancient common ancestor. **Common ancestry** is a corollary of the multiplication of species: if species arise from ancestral species multiplying, then this should produce a pattern of shared common ancestry among species, analogous to a family tree. **Gradualism**, in its most general sense, is the notion that evolution occurs in small increments over long periods of time.

Table 2.1. The five theories model of Darwinian evolution, after Mayr's *One Long Argument*.

Theory	Description
Evolution per se	This is the basic idea that lineages of organisms change over time.
Natural Selection	This is Darwin's mechanism for evolution, where individuals with traits that give them an advantage in survival and/or reproduction will be more likely to survive and produce offspring than those without the advantage.
Multiplication of Species	One consequence of evolution is that different populations of a single species could change over time until they are different enough to be considered separate species.
Common Descent	Another consequence of evolution is that multiplication of species will result in patterns of common ancestry, where species are related to each other to different degrees.
Gradualism	Evolution is a slow process that requires long periods of time to result in the kind of diversity that we see today.

We've already seen the test of gradualism in the Thomson story and the later use of radiometric dating to calculate the age of the earth. Darwin recognized that evolution was a slow process that was not obvious during the lifetime of a casual human observer and that it required a great deal of time to produce the diversity of life that we see today from, as he put it, "a few forms." Thus, Darwin's theory predicted that the earth was ancient, at least hundreds of millions of years old; current estimates of the earth's age give us an age above what Darwin's theory required at a minimum. Had radiometric dating returned results more like those of Thomson's calculations, Darwinian evolution would likely not be so thoroughly discussed in textbooks as it is today, except perhaps as an idea that was wrong. It is perhaps possible that, had Thomson been right, later biologists might have refined Darwin's ideas to accommodate a younger earth, but certainly his ideas would have been greatly altered.

Chapters 3–6 focus on the four remaining component theories, which are more integral to Darwin's ideas. While some form of Darwin's ideas might have survived the refutation of gradualism, there is no way that Darwinian evolution could have survived as an influential theory with the falsification of the mechanism that causes evolution, or the important consequences of evolution that explain the origin of species.

Further Reading

Mayr, E. 1991. *One Long Argument*, 195. Harvard University Press, Cambridge, MA.

Part II

Microevolution—Evolution of Populations

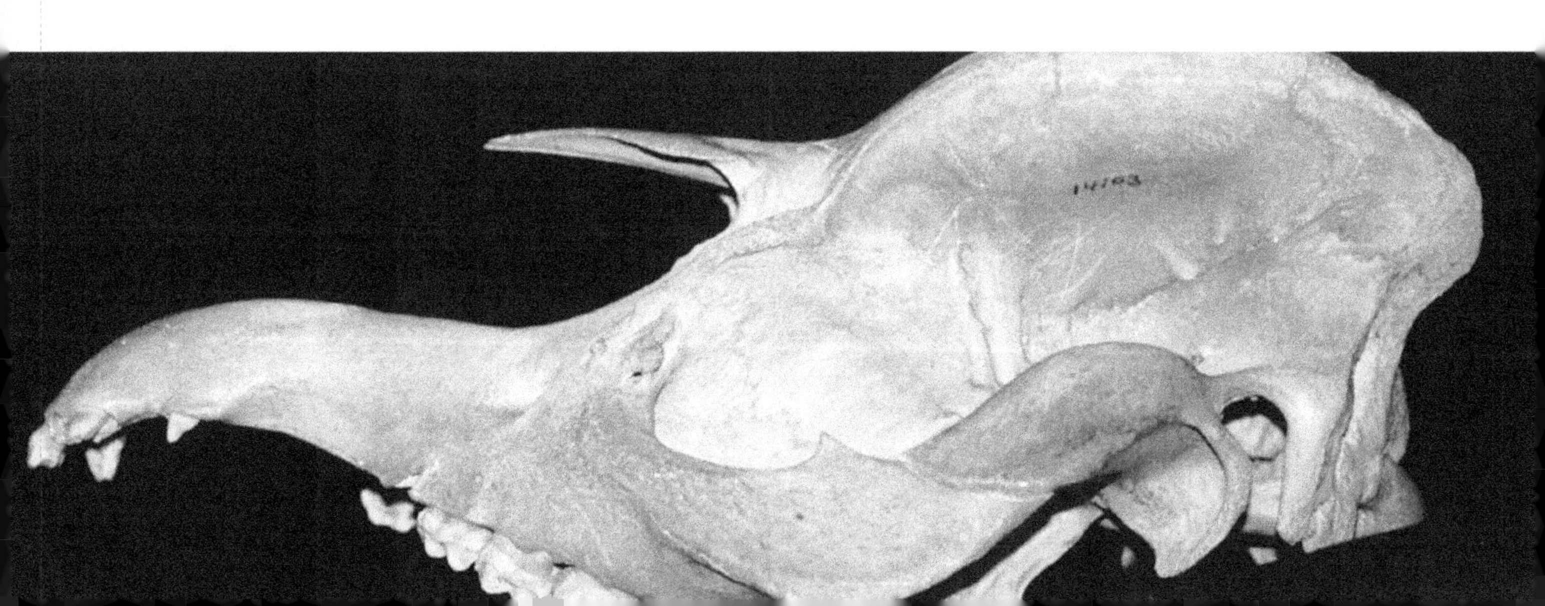

CHAPTER 3

The Logic and Testability of Natural Selection

We will begin our examination of the testability of the remainder of the five theories with Darwin's mechanism for evolution, natural selection. Natural selection is perhaps the most influential of Darwin's ideas. Like the greater theory of Darwinian evolution, we can decompose evolution by natural selection into a set of assumptions (or "facts") and inferences based on those assumptions, which form a logical argument for how these interrelated phenomena would lead to change in a population over time, or evolution. This structure provides lots of opportunities to test these assumptions and inferences, and therefore to ask whether natural selection (and evolution due to it) is really something we would expect to find in a given population, or in the natural world as a whole. Furthermore, it allows us to identify the conditions that would prevent natural selection, and to consider the specific consequences of breaking a particular link in the chain. As we shall see, there is a distinction between the process of natural selection and evolution due to it.

The logic of evolution by natural selection is summarized in Table 3.1, and this rendition is modified from Ernst Mayr's book *One Long Argument*, which we also encountered in Chapter 2.

Table 3.1. The logical structure of facts and inferences required for evolution by natural selection. After Mayr, *One Long Argument*.

Start with ...	Leads to ...	Leads to ...	Leads to ...
Fact 1: Populations tend to grow exponentially when they can.	*Inference 1: There is a struggle for existence.*	*Inference 2: Survival is not random. (Natural selection)*	*Inference 3: The frequency of heritable traits will change over time (i.e., evolution will occur).*
Fact 2: There are limits on population growth.	Fact 4: Individuals vary from one another.	Fact 5: Some variation is heritable.	
Fact 3: Populations are usually observed to stay at a steady state.			

If we want to test whether natural selection occurs, we can test the three inferences, and to test those we need to address whether the five "facts" are really just that, or whether they are poor assumptions. Even if we can verify the five facts with data, we can also test the inferences themselves to ascertain whether those facts really lead to the proposed inferences. The next three sections will address the three inferences and their related facts. After that, we will examine some further predictions of natural selection.

3.1: How population growth leads to a crowded house

The nature of population growth and its limitations was one of the most important insights that Darwin took from the economist Thomas Malthus, whom Darwin credited as being a major inspiration for natural selection. To appreciate this, we need to think about how populations grow under ideal conditions.

3.1.1. How populations grow when there are no limitations

Activity

Imagine a bacterium that reproduces by dividing into two. Start with a single such bacterium that then splits to form a new generation. Each generation thereafter, each bacterium will divide to form a new generation. No bacteria die before reproducing.

1. *Make a table showing the number of bacteria in the population for each generation for 10 generations.*
2. *Make a graph of population size versus generation.*
3. *What shape is this graph?*
4. *Can you describe the population growth with a mathematical formula?*

One thing that should be obvious from the above exercise is that the graph of population growth is not a straight line. The change in population size is slow at first, but growth rapidly increases, creating a curve. Essentially, every generation the population size doubles. We could describe this with a mathematical formula:

$$y = 2^x$$

In this case, y is the population size and x is the number of generations. Even though most populations don't grow by doubling every generation, they do grow (at least initially) in a similar fashion, producing a graph with a similar shape that can be represented with a mathematical formula that involves an exponent. We call this kind of growth **exponential growth**. Note that not every population grows as a power of two, but, when a population is experiencing exponential growth, the size of the population will be determined by a function with an exponent, where the exponent is essentially time.

Of course, our thought experiment doesn't prove that populations grow in this manner, so we should look at some examples of real populations. Figure 3.1 illustrates the initial period of growth for a yeast culture, where the population size is estimated by the amount of light absorbed by the culture over time. Another example is the reindeer of St. Paul Island: 25 reindeer were introduced to St. Paul Island off Alaska in 1911, and, after an initial few years of slow growth, by 1938 there were over 2000. The yeast graph is very similar in shape to those of the hypothetical bacteria from our last activity. But even these cases are unusual: they represent the growth of an initially small population in an environment that can accommodate expansion. Most populations that we encounter are well past that initial phase. Do they continue to grow exponentially? If not, what happens?

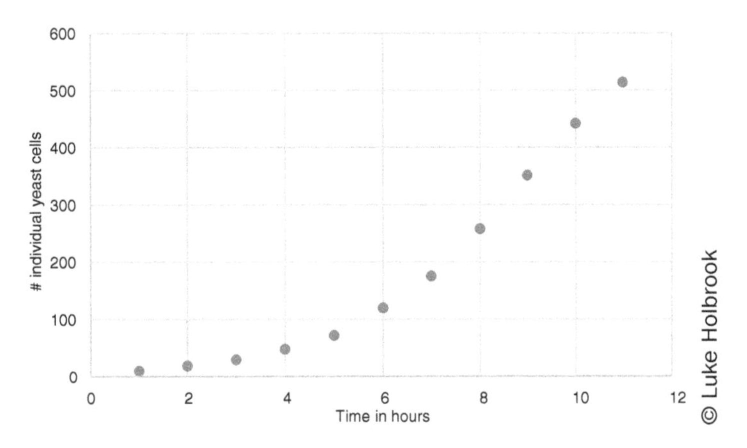

Figure 3.1. Exponential growth in yeast. Based on data from Carlson (1913).

3.1.2. How populations actually grow

Figure 3.2 illustrates the growth of the yeast colony from Figure 3.1 over a longer period of time. Note the distinctive "S"-shape to the graph; growth that shows this pattern is called **logistic growth**. We have already discussed the shape of the early, lower part of the graph, which represents exponential growth. The curving of the graph in this region from almost horizontal to almost vertical reflects how the population is adding a larger number of individuals with each successive generation. But what can we make of the later part of the graph, where it curves from nearly vertical to essentially horizontal again?

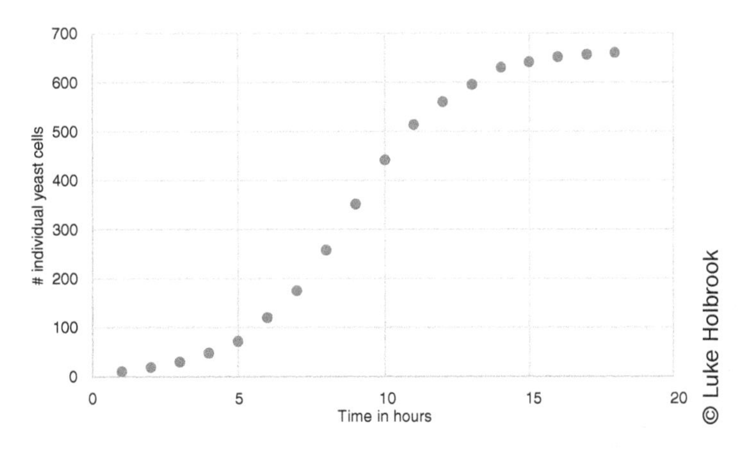

Figure 3.2. Logistic growth in yeast, extending the period of growth from Figure 3.1. Based on data from Carlson (1913).

Logistic growth essentially tacks on two notable features to the exponential growth curve. First is that it curves again, but this time in the opposite direction of how it curves during the initial period of growth. Second, the graph flattens out, essentially reaching some limit beyond which the population will not grow. This effectively tells us how many individuals the environment can support; we call this the **carrying capacity**.

Brainstorm

What we've just observed is that there are limits to how large populations can become; thus, they will not grow exponentially indefinitely. Let's now consider why population size is limited.

Working in groups with other students, come up with a list of things that will limit population size.

You probably thought of a number of things that prevent populations from growing indefinitely. You would probably come up with things like food sources (or sunlight and nutrient sources for plants), water, suitable habitats or nesting sites, and predators. Less obvious ones might include the potential for spreading communicable diseases. For something like food or nesting sites, it's fairly obvious how this would limit population: there can't be more individuals than can be sufficiently nourished by the food sources, nor can there be more individuals reproducing than there are nesting sites. Thus, it's easy for us to explain why the curve eventually becomes flat.

But why does the graph curve from the vertical to the horizontal at the top of a logistic growth curve? Why doesn't the population grow faster and faster until it hits that carrying capacity? Figure 3.3 illustrates what such a graph would look like. What, then, is the meaning of the logistic curve?

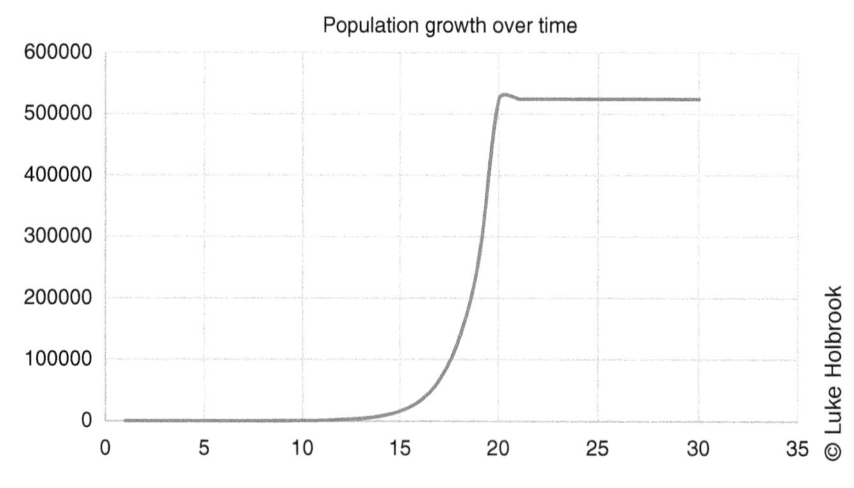

Figure 3.3. What logistic growth does not look like.

Activity

1. *Imagine you have surveyed several populations of animals and recorded how many offspring each female produced in a given year. Consider the relationship between the average number of offspring produced by a single female (y-axis) and population size (x-axis). If the population is large (or more dense), will females produce more offspring than those in a smaller, less dense population? Why or why not? Draw a graph illustrating the relationship between population size and the average number of offspring per female based on what you concluded.*
2. *Imagine you have planted seeds in pots, with different numbers of seeds in each pot. You let the seeds grow into plants over a period of time. Consider the relationship between the average weight of each individual plant in a given pot and the number of plants in that pot. Do you expect an individual plant to be heavier in a pot with more plants or in a pot with fewer plants? Why? Draw a graph illustrating the relationship between the number of individuals growing in a given pot and the average size of the individuals in that pot.*
3. *Consider your answers in terms of what is happening in a logistic growth curve as the population nears carrying capacity. As the graph flattens out, is the rate of change in population size increasing or decreasing? How do your answers to the first two parts of this activity support your answer to this question?*

What you have just investigated is the effect of **density-dependent** limitations on population growth. Density-dependent phenomena have different effects depending on the population density. Spread of disease is an example of a limit on population that is density-dependent; at low population densities, diseases don't spread very easily because infected individuals don't often encounter new individuals to infect. As the population grows, the area becomes more densely populated, and infected individuals encounter new potential hosts for the disease more often.

In fact, many limits on population growth are density-dependent, including food. It is often the case that individuals living in crowded conditions don't get as much to eat as those living in uncrowded conditions. Denser populations lead to greater competition for resources. As a result, we would predict that our hypothetical plants in pots would individually weigh less on average in pots with more plants than those in the pots with fewer plants. The competition among individual plants is greater in the pots with more plants, so it is harder for individuals to get the resources to grow larger. Figure 3.4 illustrates what this relationship would look like. We would also expect that the hypothetical animal populations you considered in Question 1 of the last activity would produce fewer young per female as the population size increases, because it is harder for females to get the energy needed to reproduce in a dense population. Figure 3.5 shows what this would look like.

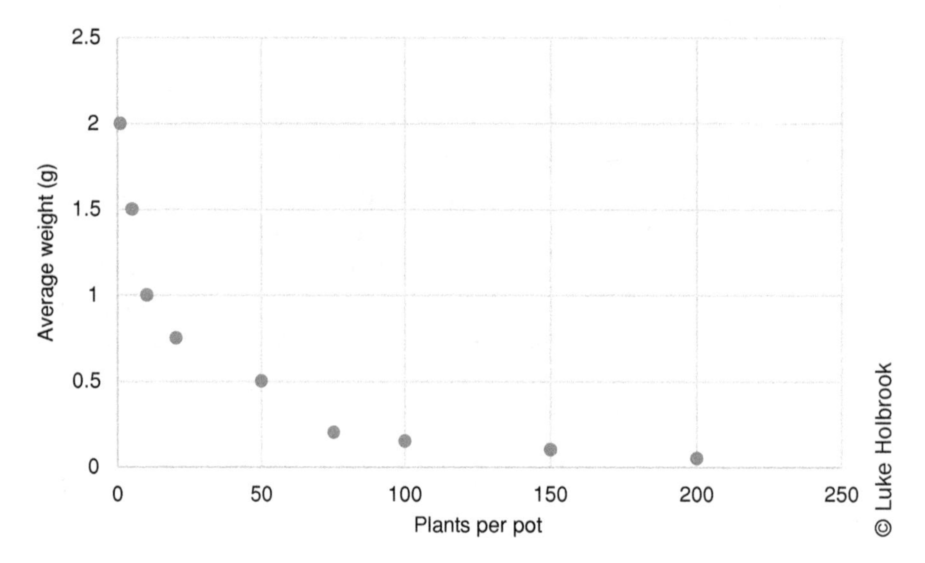

Figure 3.4. An illustration of density-dependent effects on growth, using a hypothetical example of pots with different numbers of plants.

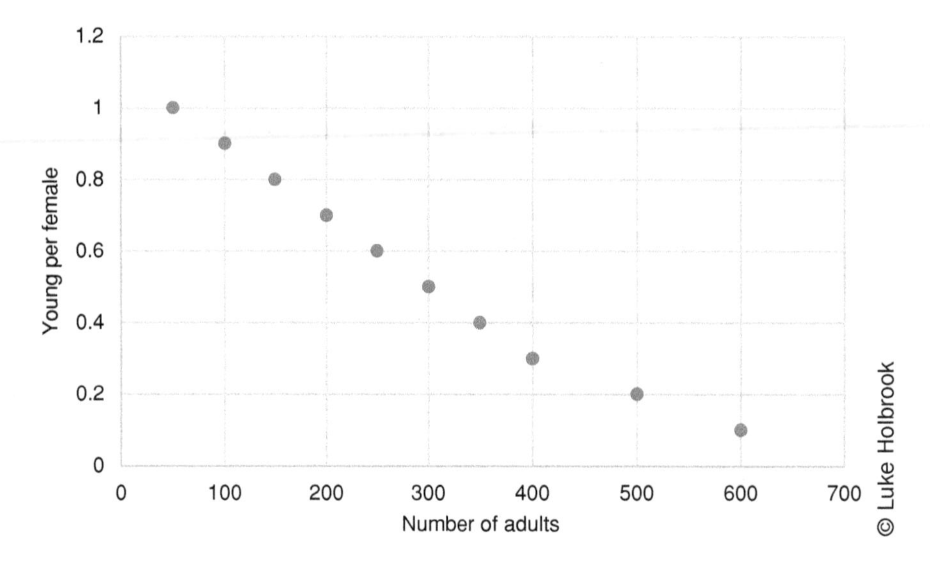

Figure 3.5. An illustration of density-dependent effects on reproduction, using a graph of a hypothetical population and the relationship between the adult population size and the number of young per female.

So, what does this tell us about what is going on at the end of a logistic curve? Just as the initial curve corresponding to exponential growth signifies an increasing rate of change in population size, the curve from vertical to horizontal signifies a decreasing rate of change in population size. Why is the rate of change in population size decreasing? It decreases because the increasing population density is resulting in greater competition for resources. As a result, individuals are finding it harder to obtain enough resources to grow, to reproduce, and even to survive.

We've established that populations grow rapidly when they can, and that they cannot grow rapidly for long before running into limitations that reduce the rate of change in population size until it is essentially zero, and the population size reaches a steady state number. Let's be clear on what is *not* causing this pattern. Individuals are not "deciding" to reduce their reproduction as the population size nears carrying capacity. Nor does the population size reach carrying capacity and everyone just stops reproducing, keeping the population size at a particular level. The steady state is reached as a result of changes in the rates of births and deaths. Individuals successfully reproduce less often because more of them are competing for the resources necessary for sustaining reproduction. Individuals die more often, because competition makes it harder for them to obtain the resources they need to survive.

Another way to state this is that the reproduction that drives exponential growth eventually produces more individuals than the environment can support. Thus, not every individual that is born survives, or at least survives to reproductive age, and not every individual that survives to reproductive age actually reproduces. There is a **struggle for existence**, both for individuals and for the offspring they have or might have.

3.2: Survival of the fittest

The struggle for existence means that not every individual in a population survives, but who survives, and who does not? For instance, is survival random, like a lottery? Or, are there other factors that determine who survives, or at least influence the chances of survival for an individual? If the former is true, then it doesn't matter whether individuals are different from one another, or whether everyone is identical. If the latter is true, then there must be differences among individuals that allow these other factors to distinguish between those more likely and those less likely to survive.

3.2.1. The nature of variation

The first thing we should therefore establish is that variation does exist among individuals. We will see examples of variation throughout this book. Scarlet tiger moths in Chapter 4 vary in the prevalence of white spots. Medium ground finches of the Galápagos, which we will discuss a little later, vary in body size and beak depth. We can demonstrate variation in a wide variety of features of a wide variety of organisms. What we should stress here is that (1) variation can occur in lots of different aspects of an organism, (2) we can measure variation in a variety of ways, and (3) variation can be the result of genetic differences, environmental differences, or (as is often the case) both. We'll address the third point in the next section, so let's say a little bit about the other two points.

Usually when we speak of variation, we think of anatomical or **morphological** variation, variation in the form or structure of the organism. We could also extend that to different levels of organization, such as variation in cell structure or the types of cells organisms have. We could even consider variation in molecular structure, such as variation in the amino acid sequences of a specific protein, like hemoglobin. Note that all of these are examples of **phenotypic variation**, because they refer to variation in observable traits. When we talk about **genetic variation**, we are specifically talking about variation in genes, and usually in the sequences of **DNA** in different individuals. Of course, when we know there is a strong relationship between genetic variation and phenotypic variation (we will encounter examples of this in the next chapter), we can use phenotypic variation as a proxy to study genetic variation.

Organisms naturally vary in many other ways besides their form. Physiological processes can vary, such as metabolic rates in different individuals. Behavior can also vary, such as differences in the songs sung by males within a species of frog. Both of these are also kinds of phenotypic variation.

3.2.2. Measuring variation

Some traits exhibit **continuous** variation. The legs of individual lizards don't simply fall into categories like long and short; they can be measured such that there are an infinite number of possible leg lengths between the length of the shortest lizard leg and the longest lizard leg, limited only by the precision of our measuring instruments. The same could be said of measuring the height of a student and his or her classmates. Other traits can be categorized and exhibit **discontinuous** variation. For instance, flower color might simply be recorded as red, blue, and so on. A species of lizard might vary in the number of rows of scales that individuals have; even though this is a number, it is always an integer for any given individual, so it is discontinuous.

When dealing with continuous variables, like a person's height, we can use some kind of instrument appropriate to such measurements, such as a meter stick. In many cases, the variation we are studying is discontinuous, such that individuals fall into one of a number of categories. One of the most obvious examples of this is looking at sex ratios, where we would categorize individuals as male or female.

Typically, when we want to study variation, we cannot study every individual in a population, so instead we try to study enough individuals such that we have a **sample** that is representative of the whole population. In order to minimize the chances that our sample is a poor reflection of variation in the population, we try to sample a lot of individuals, or, in other words, obtain a large **sample size**. We also aim for a **random** sample, as opposed to a sample that has some bias, like including only individuals of a certain age. For instance, let's say I was taking a survey to estimate what percentage of Americans have heard of DNA, and I chose my survey participants at random from the students in a college Advanced Genetics class. This would not be a very good random sample for my purpose, because the students are not representative of the American population: not only are they biology students taking a course about DNA, but they also are enrolled in college (many Americans are not or never have been), and they likely are all in their late teens or early 20s in age, whereas most adult Americans are older than that. It is easy to understand why students in a biology course might be more likely to know about DNA, but knowledge of DNA is also likely influenced by level of education (more education means more chance of learning about DNA) and age (since what you learned is partly determined by when you went to school). Thus, a random sample should avoid any bias by sampling across the spectrum of individuals in the population.

Even though we are only sampling part of the population, the number of individuals sampled will likely be large enough that we will need some way to summarize the data on individuals in order to compare one sample to another. In Chapter 1, we introduced two basic statistics: the **mean**, the average value of a measurement (say, of a particular trait) for all of the sampled individuals; and a useful statistic for understanding the variation of a trait among sampled individuals around its mean, the **standard deviation**. The standard deviation is essentially the average amount by which individuals vary from the mean. If the range of variation is wide, then the standard deviation will be large relative to the mean; if there is little variation in the sample, then the standard deviation will be small relative to the mean.

3.2.3. Who survives?

Let's return to the questions we raised at the beginning of this section regarding the result of the struggle for existence. Are the survivors determined entirely randomly, with each individual having the same chance for survival? Or are some individuals more likely to survive than others?

The most famous example of an experiment testing these questions is Bernard Kettlewell's experiments from the 1950s on peppered moths (*Biston betularia*) in Britain. The moths have two forms, one light and mottled (Figure 3.6) and one very dark. At the time of Kettlewell's experiments, rural areas were relatively free of pollution, and the trees in these areas tended to have lighter bark, often covered with light-colored lichens. Light colored moths are very effectively camouflaged, or **cryptic**, against such a background, whereas dark moths are very conspicuous. In industrial areas, soot tended to kill off the lichen and blacken the bark, and in such areas the dark moths are hard to distinguish from the bark, whereas light moths are conspicuous.

Marek R. Swadzba/Shutterstock.com

Figure 3.6. Peppered moth (*Biston betularia*) illustrating the light morph.

Activity

1. *Come up with a hypothesis regarding how the color morphs will affect the survival of individual moths.*
2. *State the null hypothesis for the alternate hypothesis you proposed for #1.*

Your hypothesis was probably something along the lines that the color of the moth would affect its survival. Conspicuous moths would be more likely to be detected by predators, and camouflaged moths would avoid predation more easily. Thus, we would expect greater survival of light moths versus dark moths in the unpolluted woods and greater survival of dark moths versus light moths in polluted woods. The null hypothesis would be that color has no effect on moth survival.

To test this, Kettlewell marked and recaptured moths in polluted industrial areas and unpolluted rural areas. This means he caught a sample of moths, making sure to catch a substantial number of each color morph, and marked them in a way that he could identify them if he caught them again later. After some time, he captured moths again and recorded how many of the marked moths he recovered. Note that Kettlewell would not expect to recapture every moth he marked; rather, he would capture some percentage of the marked moths, in part because some moths would have died, but largely because he was not capturing every moth in the population. Even though he did not recapture every marked moth, we could still make inferences based on relative comparisons. For instance, did he recapture a larger percentage of the light moths he marked versus the dark ones, or vice versa?

Activity

1. *Based on the alternate hypothesis you generated in #1 of the previous activity, what do you predict Kettlewell's results should look like for a polluted wood and for an unpolluted wood, in terms of the percentages of marked light and dark moths he would recover?*
2. *Based on the null hypothesis you stated in #2 of the previous activity, what do you predict Kettlewell's results should look like for a polluted wood and for an unpolluted wood?*
3. *Examine the results of Kettlewell's experiments in Table 3.2 and evaluate the alternate and null hypotheses based on these data.*

Table 3.2. Results of Kettlewell's (1955) experiments on peppered moths (*Biston betularia*).

	Unpolluted wood			Polluted wood		
	Released	**Recaptured**	**Percentage recaptured**	**Released**	**Recaptured**	**Percentage recaptured**
Melanic	406	19	4.7	447	123	27.5
Typical	393	54	13.7	137	18	13.1

Our predictions based on our alternate hypothesis would be that we expect a higher percentage of cryptic moths to be recaptured than conspicuous moths. Thus, in the unpolluted woods, we expect a higher percentage of light moths to be recaptured than of dark moths, and in the polluted woods we expect the opposite, that a higher percentage of dark moths will be recaptured than of light moths. The null hypothesis predicts that there should be no significant difference in percentages recaptured for either morph in either type of wood.

The results of Kettlewell's experiments match the predictions of the alternate hypothesis. We can even use the percentages to quantify how much of a benefit there is to being the right colored moth (or the cost of being the wrong colored moth); in other words, we can measure **fitness** and the **strength of selection**.

Let's start with fitness, a term we use to describe how likely individuals are to survive, or, more accurately, how likely they are to pass their genes to the next generation, presumably by surviving and reproducing. In this case, we cannot easily make an absolute measure of fitness, but we can make a measurement of **relative fitness**. To calculate relative fitness (ω), we need some measure related, at least in this case, to survivorship. We can use the percentage of moths recaptured for each type in each wood. By dividing that number by the highest percentage, we effectively set the relative fitness of the fittest form at 1.0, and the relative fitness of less fit forms will be some number less than 1.0.

Activity

1. *Calculate the relative fitnesses of the light and dark moths for the unpolluted and polluted woods. Note that you need to treat the two types of woods separately.*
2. *Which form had a relative fitness of 1.0 in the unpolluted woods? Which had a relative fitness of 1.0 in the polluted woods? What were the relative fitnesses of the less fit forms in each type of woods?*

You presumably noticed that light moths in the unpolluted woods and dark moths in the polluted woods had relative fitnesses of 1.0. The dark moths in the unpolluted woods had a relative fitness of 0.34, and the light moths in the polluted woods had a relative fitness of 0.48. One thing that you should notice is that selection appears to be acting to different degrees in the two different types of woods; in other words, there is stronger selection in one of these woods than the other.

One way to represent this is by calculating the strength of selection (s). We calculate s by subtracting the lower relative fitness from the higher; in other words:

$$s = 1 - \omega$$

The higher s is, the stronger selection is acting on individuals.

Activity

1. *Calculate the strength of selection for the unpolluted and polluted woods.*
2. *In which wood is selection acting more strongly?*

What can we learn from the strength of selection? For one thing, we can infer that selection will drive evolution toward a particular result more quickly when selection is very strong. So, because selection is stronger in the unpolluted wood than in the polluted wood, we might expect the frequency of light moths to increase faster in an unpolluted wood than the frequency of dark moths would increase in a polluted wood. Note that we haven't determined the exact reason why selection is stronger in the unpolluted wood. We could propose specific hypotheses for this that would require another set of experiments to test: for instance, perhaps birds have an easier time noticing a dark object on a light field than a light object on a dark field; or perhaps producing the dark pigment costs the moth more energy or has some other cost. In any case, we'll leave such questions and continue discussing the logic and testing of natural selection.

To this point, we have established the occurrence of **natural selection**, that is, that survival (and reproduction) is not random but is influenced by variation among individuals. In other words, we've established that the nature of population growth and the nature of variation results in a struggle for existence where the survivors (or those who successfully reproduce) are those that have traits that confer the greatest fitness. Natural selection is sometimes defined as "differential reproduction," because it results in differences between individuals or kinds of individuals in their ability to contribute offspring to the next generation. It is common to think of natural selection in terms of survival—after all, individuals that do not survive also do not reproduce. We should, however, keep in mind that individuals can vary in terms of reproductive success for reasons that have little to do with survival. We will explore some examples of this later.

In order for natural selection to result in evolution, we need one more ingredient, namely that these traits conferring fitness are passed on to offspring.

3.3: Heritability, or how natural selection becomes evolution

The last piece of Darwin's logic for natural selection as a mechanism for evolution is that, if the traits that determine survival are inherited by the offspring of the survivors, over time the population will change in terms of the variation observed in this trait in the population. In other words, more members of the population will exhibit forms of the trait that increase survival, and they will produce more offspring with those traits, who will be more likely to survive and produce more offspring of their own with these traits, and so on. But how do we know that traits are heritable?

3.3.1. Measuring heritability

If the traits of individuals are determined by their parents, then we can make predictions about the relationship between offspring traits and parental traits. The obvious prediction is that offspring and parents should be similar to one another. But how similar do they need to be for us to accept that the trait is heritable? How can we be sure that they are not similar just by coincidence? One statistical tool we can use to test this is **correlation**.

A Closer Look: Correlation

When we say two variables are correlated, we are saying that we can predict the value of one variable by knowing the value of the other. Sometimes, we have reason to believe that one variable is really determining the value of the other; this is what we call **causal correlation**. But it's not always easy to infer a cause from a correlation. Cholesterol levels may be correlated with risk of heart disease, but is that because high cholesterol causes heart disease, or is it because something else that we haven't identified causes both? This is what is meant when people warn not to confuse causation and correlation. Nevertheless, we can often make predictions about correlation to test hypotheses about causal relationships, and we can do that to test heritability. As a result, when we look for correlations, we usually have identified one variable that we think is determined by the other. The determining variable is called the **independent variable**, whereas the variable whose value is determined by the independent variable is called the **dependent variable**.

The simplest way to represent a correlation is to plot a graph of two different variables that have been measured. Figure 3.7 illustrates three different plots of data examining a correlation. In the first graph, all of the points fall perfectly on a line. In fact, we can fit a line to the data and describe it mathematically with a slope and y-intercept. The relationship between the variables indicates that if I know the value of the independent variable, I can predict with perfect precision the value of the dependent variable. In this case, a low value for the independent variable predicts a low value for the dependent value, and a high value for the independent variable predicts a high value for the dependent variable. This kind of correlation is called a **positive correlation**.

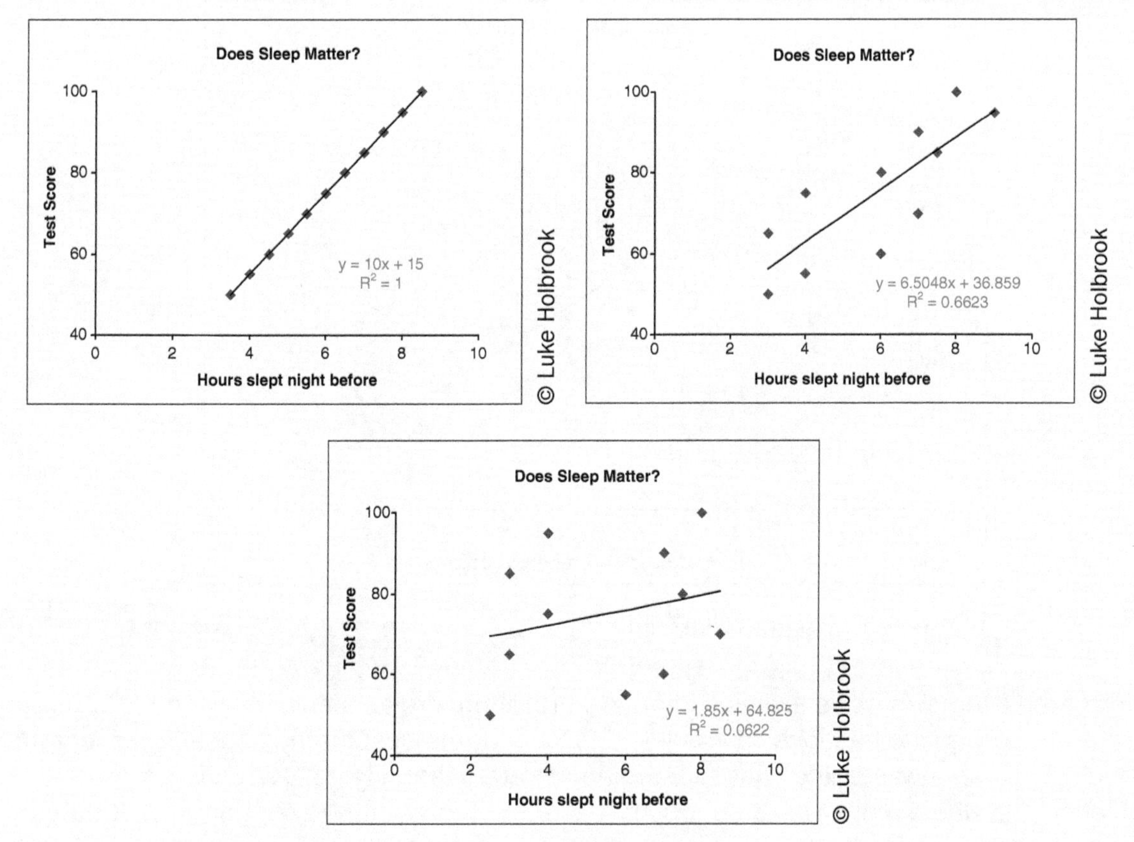

Figure 3.7. Perfect, good, and poor positive correlations.

Perfect correlations are uncommon in biology, and the other two graphs in Figure 3.7 illustrate more common occurrences with biological data. The second graph illustrates data with greater "spread" to them than in the first graph. The points don't all fall along a line, but there is enough of a trend in the data that the independent variable gives us some power to predict the dependent variable. We can fit a line to the data that approximates the trend. In fact, we can use statistical methods to fit a line for us, and we can also calculate a statistic called **Pearson's correlation coefficient** (*r*) that tells us how much correlation exists between the two variables.

The *r* value essentially evaluates how the one variable changes when the other does. The *r* value is an index, which means it has a value between −1 and 1. The perfect fit of the line to the data in the first graph produces a value for *r* of 1. The worst fit of the line to the data would produce a value of 0. The third graph in Figure 3.7 comes close to this, where $r = 0.0622$. The second graph falls in the middle: the fit for that graph is $r = 0.6623$.

You may be wondering why the *r* value can be as low as −1, when the worst fit gives a value of 0. Not all correlations are positive. A **negative correlation** occurs when the relationship between the variables is inverted: a high value for the independent variable determines a low value for the dependent variable, and a low value for the independent variable determines a high value for the dependent variable. A perfect negative correlation would give a value for *r* of −1. The graph in Figure 3.8 illustrates a hypothetical example of a fairly strong negative correlation, where $r = -0.8819$.

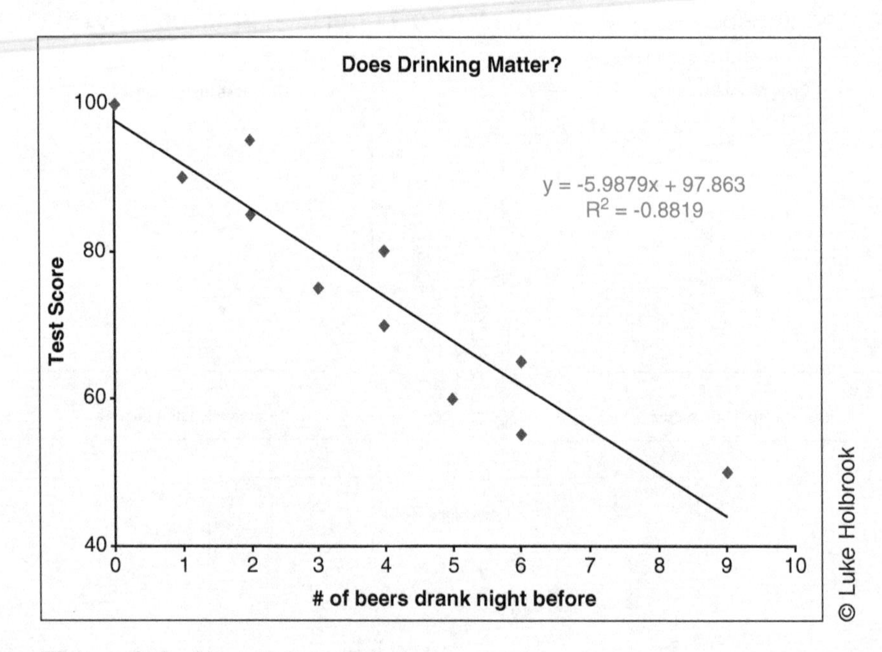

Figure 3.8. Negative correlation.

All of the examples we've examined not only predicted a relationship between the two variables, but also that the relationship was linear, such that the best mathematical representation of the relationship was a line. Not all correlations are linear. Figure 3.9 gives an example of a relationship where a line would be a poor device for describing the relationship, but a curve performs almost perfectly. Note that we can still calculate a value for *r*, which is only 0.0158 for the line but almost perfect at 0.9845 for the curve.

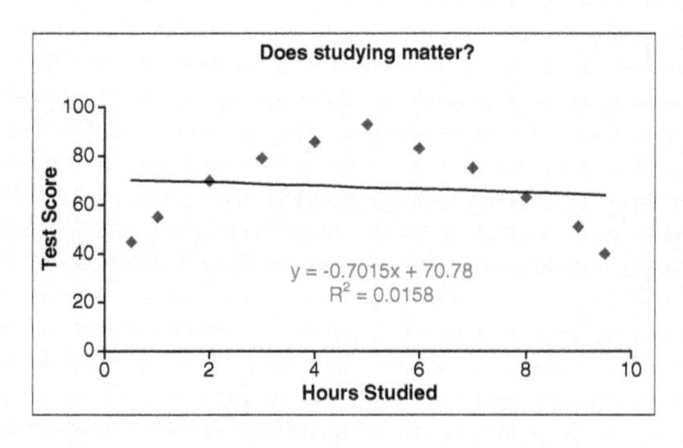

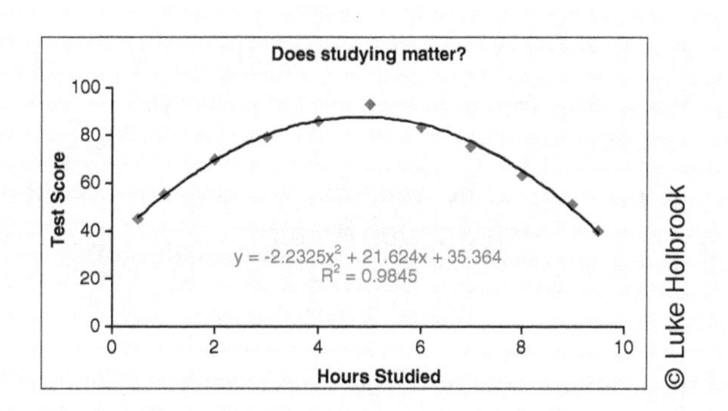

Figure 3.9. Nonlinear correlation.

Let's now ask what kind of correlation we would predict for a trait that is completely heritable. The answer is a positive correlation, and in fact we'd expect a slope of 1.0 and a value for r of 1.0, since offspring should have exactly the same condition as their parents. Of course, as we consider plotting the data, there is one issue that immediately presents itself: each offspring has two parents. For a continuously varying trait, like beak depth, one way we could get around this is by averaging the values for the two parents and plotting that against the value for an individual offspring.

Activity

Figure 3.10 shows data gathered on beak depths of parents and offspring of medium ground finches on Daphne Major in the Galápagos for 1976 and 1978.

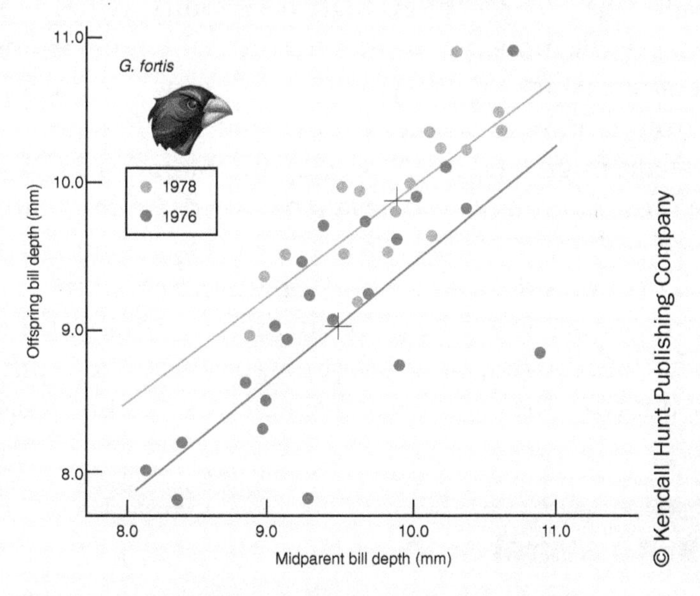

Figure 3.10. Data on beak depth from medium ground finches on Daphne Major recorded for 1976 and 1978.

For both years, the r value for the line of best fit is about 0.9.

1. *Does the information from the graph support the hypothesis that beak depth is heritable? Why or why not?*
2. *Is beak depth heritable in one year but not the other? Explain your answer.*
3. *Is beak depth a variable trait?*

The answer to the first question is that the data do support the hypothesis that beak depth is heritable. It's possible that you might have answered "no" if you based your answer on the predictions for a trait with complete heritability, because the r value is not exactly 1.0, nor is the slope exactly 1.0. Of course, there are other reasons why a heritable trait might not meet these predictions exactly. The main one is that many traits are influenced both by heredity and by the environment. So, we could explain the deviations from the line of best fit as the influence of environmental effects on individuals.

In some cases, we can even control for these environmental effects. Smith and Dhondt (1980) placed eggs from one song sparrow's nest into another and compared the beak depth of the offspring with those of their biological and foster parents. If the trait is heritable, we would predict that there would be a correlation between offspring and biological parents, but not between offspring and foster parents. This in fact is what they found.

The answer to the second question is that the trait is heritable in both years. If a trait is heritable to any degree, that presumably is because the trait is partly determined by the genes the individual carries. If that is true for one generation, it should be true for other generations as well, so we don't expect heritability to change from generation to generation. Of course, the lines on the graph for both years match what we would predict for beak depth being heritable. Note also that in order to evaluate heritability we need to consider each generation separately, because the individual values for different generations could be different because natural selection was acting differently in those two years. The data represented in Figure 3.10 come from landmark studies led by Peter and Rosemary Grant, who tagged and measured every medium ground finch on Daphne Major in 1976 and 1978. Due to a drought in 1977, the most abundant seeds available for these seed-eating birds were large and hard-shelled, favoring birds with deeper beaks to crack open the seeds. As the graph shows, the Grants observed that the average beak depth increased from 1976 to 1978.

The answer to the third question is, yes, beak depth is a variable trait. Note that there is a range of beak sizes both for parents and offspring; parent beak depth largely determines offspring beak depth, but that does not mean that all birds hatched in a given year will have the same beak depth.

3.4: The effects of natural selection

The examples we've looked at so far give us some idea of how natural selection will change a population over time. In the case of moths in a particular wood, we expect that one color morph will be favored, and eventually the whole population might exhibit that one phenotype. In the case of beak depth of Galápagos finches, if the dry conditions continued, we might expect beaks to get deeper and deeper until reaching some limit. (They certainly can't be deeper than the birds' heads!) As it turns out, the drought on Daphne Major was followed a few years later by an unusually wet period, and selection under these conditions actually favored smaller birds with smaller beaks.

In the cases of the finches and the moths, the direction of change due to selection depended on the environment. In all cases, though, selection changed phenotypes observed in the population in a predictable direction. In the case of the finches, the observed effect of selection was to change the mean value for the trait in the population over time, without necessarily changing the amount of variation (which we could measure with the standard deviation). We might also consider if there are other patterns that natural selection would produce.

Activity

For each of the following cases, draw the indicated graph illustrating how selection will affect the population. Specifically, state how the mean value and standard deviation for the trait will be affected for each case.

1. *Trophy hunting for bighorn sheep (*Ovis canadensis*) tends to target the largest males and males with the largest horns. Predict how mean body mass and mean horn length will change over time in bighorn sheep populations and illustrate this with graphs. Compare your predictions with the results in Figure 3.11.*

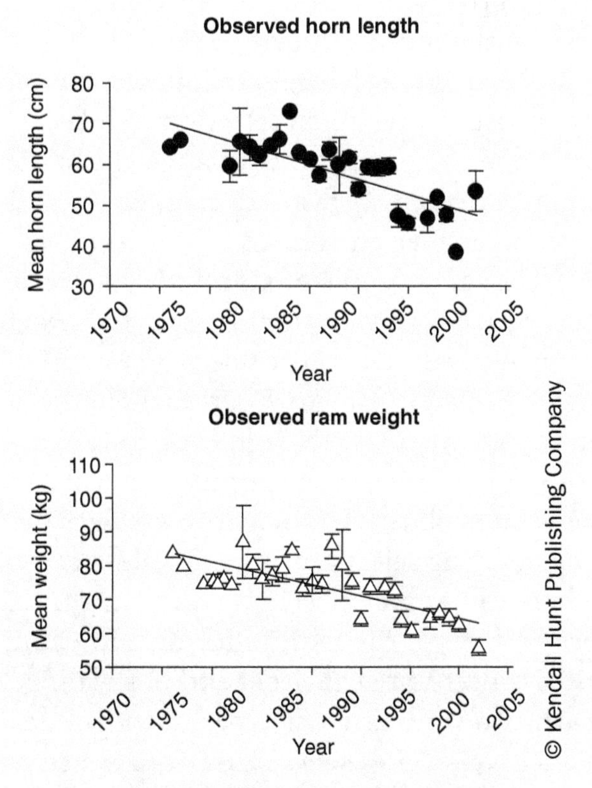

Figure 3.11. Selection on horn length and ram size due to hunting of bighorn sheep. From Coltman et al. (2003).

2. *Different species of birds vary in the number of eggs they produce in a clutch, but the variation in clutch size within a species is usually fairly narrow. For any given species of bird, more eggs laid should increase fitness, but, if more eggs are laid than the parents can feed after hatching, too many eggs could result in decreasing fitness. How do you predict selection will act on variation in clutch size in a population of birds? Illustrate this with a graph of the distribution of different hypothetical clutch sizes before selection and later after selection has had some time to operate.*
3. *Coho salmon (*Onchorhynchus kisutch*), like other salmon species, spawn in freshwater streams. Males often fight each other for access to females, and larger males generally have an advantage in these conflicts. However, smaller males called "jacks" can successfully pursue a strategy of "sneaking," whereby they fertilize a female's eggs without fighting for access but by hiding near a female's nest site. Smaller males are generally better able to execute this sneaking strategy. Intermediate males would be at a disadvantage for both fighting and sneaking strategies. Predict how selection for these two strategies should affect the distribution of male body size in the population and draw a graph of what this would look like before and after selection.*
4. *For each of these cases, how will the mean for the trait change, if at all? How will the standard deviation for the trait change, if at all?*

The cases above illustrate each of the three **modes of selection**. The case of the bighorn sheep illustrates the most familiar type of selection, **directional selection**, which is the same mode that we have observed in the finches and peppered moths. In this mode, the general effect of selection is to change the mean value of the trait being selected in one direction. Note that directional selection can go in either direction; selection for large beaks and selection for small beaks both constitute directional selection. While directional selection results in a change in the mean, it does not necessarily change the standard deviation; in other words, it doesn't have to change the range of variation in, say, beak depth, even if beak depth has evolved to become larger or smaller on average.

The case of bird clutch size is an example of **stabilizing selection**. In this case, extreme values at both ends of the spectrum of variation are selected against, and selection favors individuals close to the mean. As a result, because selection essentially favors the mean value, the mean for the population does not change. What does change is the range of variation, which becomes narrower as individuals with higher or lower values are selected against. Thus, the standard deviation for the trait in the population will decrease over time.

Finally, the case of the salmon illustrates **disruptive selection**. Disruptive selection is essentially the opposite of stabilizing selection: extreme values at both ends are favored, and individuals with intermediate values are selected against. In this case, individuals that are larger or smaller are favored. The result is a **bimodal distribution**, where the distribution of the variation in the trait is concentrated at either end of the range of variation. Often, this results in a population with two very different types of individuals; this is why disruptive selection is sometimes called **diversifying selection**. Because after selection there are lots of individuals at both ends of the range of variation, the mean value of the trait for the population does not necessarily change. But, because selection results in so many individuals that have values for the trait that are far from the mean, the standard deviation increases over time.

3.5: Sexual selection, a type of selection not based on survival

As noted before, natural selection is more than "survival of the fittest," and in fact we define fitness not by survival, but by an individual's ability to pass its genes on to the next generation. Thus, an individual that lives a short life but still manages to produce many offspring would have a high fitness, whereas a long-lived individual that never reproduces has a fitness of zero. Selection therefore can act on numerous traits that are unrelated to survival, including differences in the number of offspring produced, differences in how individuals allocate resources to reproduction versus growth and survival, differences in how much parental care is provided, and so on.

In sexually reproducing species, an important part of fitness is mating success: an individual needs to contribute to producing a fertilized egg with another individual in order to produce offspring. Thus, selection can act on traits that influence mating success, and it might act differently on males and females. For instance, Jones, Arguello, and Arnold (2002) examined differences in mating and reproductive success of male and female rough-skinned newts (*Taricha torosa*). In this species, every female has at least one mate and produces some offspring, with females typically having two or more mates and producing 150 or more offspring. The vast majority of males, on the other hand, never mate and, consequently, produce no offspring; only a small number of males are responsible for all of the matings. Thus, the vast majority of males of this species have a fitness of zero, and this is largely due to a failure to obtain a mate. Any trait that increased a male's chances of mating, therefore, would have a dramatic effect on fitness and would be heavily favored by selection.

Sexual selection is natural selection based on traits related to mating success. Darwin came up with this concept and made it the central theme of one of his most important works, *The Descent of Man*. Sexual selection, as we will see, helps to explain differences that we observe between males and females. To understand why sexual selection happens, and why it tends to take certain forms, it is helpful to think about the effective reproductive strategies entailed by almost all males and females.

3.5.1. Male and female reproductive strategies

In multicellular organisms, sexual reproduction involves the combination of two sex cells, or **gametes**, typically from two different individuals. The gametes are **haploid**, meaning they contain one chromosome from each pair of chromosomes that make up the individual's **diploid** genome. Combining the two haploid gametes forms a new diploid individual. In some cases, the gametes forming a new individual are cells that look similar (**homogametic**), but far more commonly we see gametes that come in two forms, each of which is necessary for forming a new individual (**heterogametic**). These are familiar to us as **sperm** and **eggs**. In both plants and animals, sperm are small, often mobile, and produced in large numbers; the pollen that coats my car in the spring and that triggers allergies is essentially innumerable particles, each containing plant sperm. Eggs, on the other hand, are some of the largest cells, and, while they can be produced in large quantities, their numbers pale in comparison to those of sperm.

Sperm and eggs represent different reproductive strategies that ultimately determine the mating strategies of males and females, respectively. In producing sperm, males are investing minimal energy into each individual gamete and instead maximizing the number of gametes they produce, in an effort to ensure that at least some of their genes will end up in a fertilized egg that survives to reproduce. Females, on the other hand, are making greater investments into each gamete, as a way of ensuring that, once fertilized, each egg will have a greater chance of enduring until it develops into a reproductive individual.

These two strategies have important consequences for mating success. Because males produce so many sperm, it is possible for a single male to fertilize many eggs, and to fertilize the eggs of many females. This is apparent in the newts: a small number of males fertilize the eggs of all of the females. No female salamander lacks for an opportunity to have her eggs fertilized, but which male fertilizes her eggs should be important to her, from the standpoint of selection. She has invested far more energy into each of her eggs than the male has into each of his sperm, and yet the fitness of her offspring will be determined in part by the genes of the male that fertilizes her eggs. Thus, any traits that make her more likely to mate with a male whose genes will increase the fitness of her offspring will be favored.

3.5.2. Sexual dimorphism

The main evolutionary question that sexual selection helps to answer is why we observe differences between males and females, a phenomenon known as **sexual dimorphism**. Based on what we just noted about the different reproductive strategies of males and females above, the most obvious difference between the two sexes is that females produce eggs and males produce sperm, and as a result they possess different structures (typically called gonads) for producing these different cells. They might also possess different structures, such as genitalia, that are related to facilitating the union of sperm and eggs in fertilization. These differences among sexes that relate directly to reproduction are called **primary sexual dimorphism**.

It is worthwhile to note that not all sexually reproducing species have distinct males and females. Some species are **hermaphroditic**: individual earthworms, for example, possess both egg-producing ovaries and sperm-producing testes. Many flowering plants have flowers that hold both male and female structures, while others might have distinct male and female flowers that can be present on the same individual plant. Some species, such as certain fishes and some mollusks, might start life producing one type of gamete and switch later in life to producing the other; such species are referred to as **sequential hermaphrodites**.

Even in the case of sequential hermaphrodites, there are often obvious differences between females and males that go beyond structures directly related to reproduction (Figure 3.12). Males and females might differ in size, coloration, anatomy, physiology, or behavior. These differences that are not directly relatable to reproduction are referred to as **secondary sexual dimorphism**. Some of these differences might be related to differences in the ecology of males and females, but others cannot be explained in those terms, especially features that could be construed as reducing fitness. Surely, doesn't the exaggerated tail of a peacock hinder its ability to fly and make it harder for it to avoid predators? These kinds of paradoxical traits were the reason Darwin came up with sexual selection, because such traits could increase fitness if they increased mating success.

Petr Simon/Shutterstock.com

Dmitry Demkin/Shutterstock.com

Creativex/Shutterstock.com

Figure 3.12. Examples of secondary sexual dimorphism. (A) A male and female pheasant (*Phasianus colchicus*). The male is the brightly colored one. (B) A male and female red deer (*Cervus elaphus*). Note the antlers on the male. (C) A male (left) and female elephant seal (*Mirounga angustirostris*). The male is much larger and has a proboscis that it can inflate as a display during territorial conflict with other males.

3.5.3. Male–male competition

Many of the features that come to mind when we think of secondary sexual dimorphism are distinctive features of males: the peacock's tail, the mane of a male lion, the antlers of a stag, the bright red color of a male Northern Cardinal, and the songs of male songbirds and male frogs are some familiar examples. Based on the male reproductive strategy that allows one male to fertilize many females, as observed in male rough-skinned newts, we might expect selection to be strong for male traits that increase mating success. In some cases, these are traits that affect the competition between males for access to mates.

Can we demonstrate that such traits are actually important for male mating success by determining which males get access to females for mating? A good example comes from deer, mammals in the family Cervidae, which are distinguished from other mammals by their antlers that they shed and regrow annually. In most deer, only males have antlers, but caribou (or reindeer; *Rangifer tarandus*; Figure 3.13) are an exception where both sexes have antlers, though even here the antlers of males are generally larger. Like other deer, male caribou use their antlers in sparring contests during the mating season, and the winners of these contests gain access to a group of females for mating.

Jukka Jantunen/Shutterstock.com

Figure 3.13. A male caribou (*Rangifer tarandus*). Both sexes have antlers, but those of males are generally longer.

Do the antlers of males affect their mating success? A study by Barrette and Vandal (1990) recorded the results of 713 sparring contests between males with antlers of unequal lengths. The male with smaller antlers gave up first in about 90% of the contests, strongly suggesting that antler length is important for determining success in these contests.

The traits determining success in male-male competition for mates are not always related to "weapons" or physical strength. Male red-winged blackbirds (*Agelaius phoeniceus*; Figure 3.14) have bright red patches on the upper wings (called epaulets) that are not found in females. Males patrol a territory, displaying their epaulets and making their distinctive calls. The purpose of these displays, however, is not to attract females to their territory, but to dissuade other males from entering it.

Scott palmeri/Shutterstock.com

Figure 3.14. A male red-winged blackbird, showing its red epaulets.

Do the red epaulets affect a male's ability to maintain control of a territory? Peek (1972) modified the epaulets of some males, making them darker, and compared the number of intrusions by other males on the territories of darkened birds with those of undarkened birds. Peek's results showed that other males would trespass into the territories of males with darkened epaulets far more than they would trespass into territories of males with undarkened epaulets.

3.5.4. Female choice

While some sexually selected traits might be important only for a male's interactions with other males, others might increase a male's chances of attracting a female. Such traits are most apparent in species that have mating systems where males engage is some sort of display to attract females to them. Peacocks fan their lavish tails, male songbirds sing, male frogs call, and other species, like sage grouse, have males that perform elaborate dances, and a female will go to and mate with a particular male, presumably based on this display.

Can we demonstrate that females choose mates on the basis of a particular male trait? A good example of a study addressing this question is Møller's (1988) study of barn swallows. Like other swallows, barn swallows have v-shaped tails, and the sides of the tail are generally longer in males than in females (Figure 3.15). Møller clipped the tail feathers of some males in two places between the tip and the base, essentially creating a middle piece that he removed from some of the males, creating shorter tails. He could also take those removed pieces and create an artificially long tail by inserting that piece into the clipped tail of another male. He compared these shortened and elongated treatments to two controls, one that was not clipped at all, and one that he clipped like the others but then glued the pieces back together so they were essentially the same length as before clipping.

Thanakorn Hongphan/Shutterstock.com

Figure 3.15. Barn swallows showing variation in tail length.

Møller then observed the mating and reproductive success of the males in each treatment. He measured two dependent variables: premating period, essentially the amount of time that passed before a male mated; and the number of fledglings—offspring that lived long enough to leave the nest—each male produced.

Activity

1. *State the hypothesis Møller is testing and the null hypothesis.*
2. *Which of the two variables that Møller measured is related to mating success? Which one is related to reproductive success?*
3. *Why did Møller use two control groups?*
4. *Come up with predictions of how premating period should compare among the four treatment groups if the alternate hypothesis is correct. How should they compare if the null hypothesis is correct? Do the same for the alternate and null hypothesis with regards to number of fledglings.*

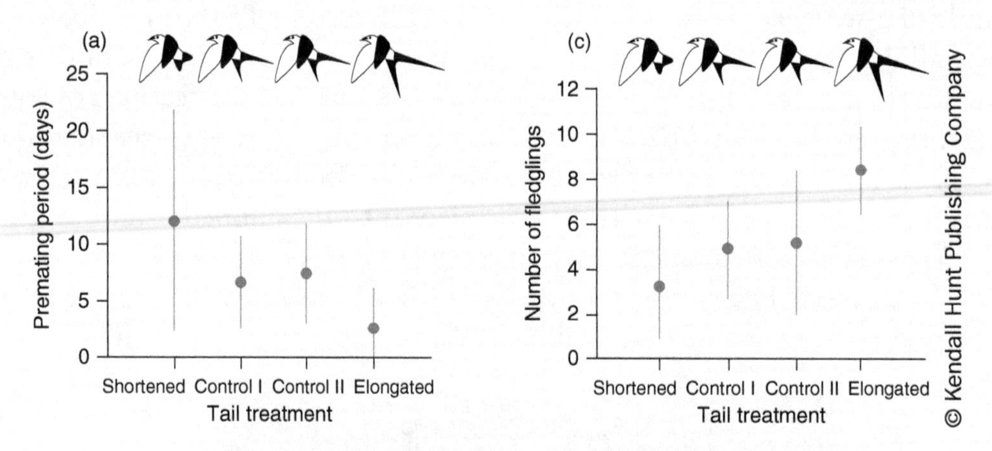

Figure 3.16. Tail length versus premating period, number of fledglings

5. *Figure 3.16 shows Møller's results. How do they compare to the predictions you made above?*

3.5.5. Why do females choose certain traits?

Hopefully, you recognized that Møller's results matched what we would expect if females preferred to mate with males with longer tails. But why do females prefer this trait? There are two hypotheses for why a given trait is preferred by females.

On hypothesis is what is called **runaway selection**. In essence, it says that the only reason the trait increases male fitness is the fact that females prefer it. To understand how this might arise, imagine a population where males vary in a given trait, let's say tail length, and most females have no particular preference for males with a certain tail length, but some females happen to prefer males with long tails. Males with long tails will have greater mating success, because a part of the female population prefers to mate with them. If tail length and preference for long tails are both heritable, then these males and the females who prefer them will have sons with long tails and daughters who prefer long tails, creating a self-reinforcing pattern of mating preference. Over time, males with long tails and females that prefer long tails will increase in frequency in the population, until it is essentially a population of long-tailed males and long-tail-preferring females.

The second hypothesis makes a more testable claim, namely that the trait preferred by females is linked to some other trait of the male that affects the fitness of their offspring. In other words, traits like long tails could be advertising the good genes that a male has. Let's return to the example of barn swallows. To test whether males with longer tails had better genes than males with shorter tales, Møller (1990) measured tail lengths for males and then counted the number of mites that their offspring had. They also swapped eggs between nests, so that some of a male's eggs were raised in a nest with a different father.

Activity

1. *State the hypothesis and the null hypothesis being tested.*
2. *Why were some eggs placed in nests that were not where the biological father resided?*
3. *Come up with predictions based on the alternate hypothesis of how a father's tail length and the number of mites on his offspring should compare, both in nests where the father resided and in "foster" nests. How should they compare if the null hypothesis is correct?*

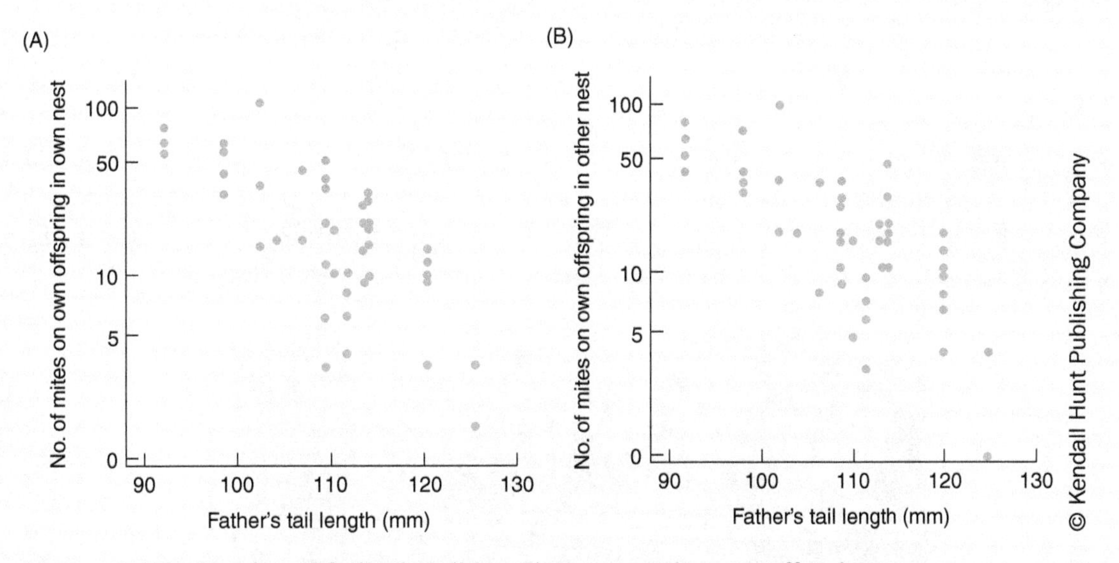

Figure 3.17. Graphs of father's tail length versus mites on offspring

4. *Figure 3.17 shows the results of this study. How do they compare to the predictions you made above?*

As is hopefully evident, there is some support for males with longer tails carrying genes that confer some resistance to mites, and therefore females who choose males with long tails are choosing males with good genes for their offspring. Why might we expect this to be the case more generally?

Many of the traits that we expect to be sexually selected based on female choice are distinctive because they are costly. A peacock's tail is costly in (at least) two ways. First, it costs energy to create and maintain a large, elaborate tail. Second, there is a potential cost in terms of fitness, as a long tail is an impediment to avoiding predators. As a thought experiment, let's divide a hypothetical population of peacocks into those with long tails and those with short tails. Within those groups, we can have males with good genes or with bad genes. Because the tail is so costly, a male with a long tail and bad genes is unlikely to survive. On the other hand, a male with a short tail and bad genes might be able to survive just as well as a short-tailed male with good genes, because neither has sunk energy into producing a long tail. This means that if a female encounters a long-tailed male, it is unlikely that it will be a male with bad genes, but a given short-tailed male might be equally likely to have good or bad genes. Therefore, females that prefer long-tailed males are making a better bet for their offspring, and selection will favor females who prefer long-tailed males.

3.5.6. Sexual selection with sex-role reversal

Sexual selection, as we observed earlier, is a consequence of the different reproductive strategies of males and females, specifically of how they invest in individual gametes. In some species, the relative investment of males and females is reversed. For instance, seahorses and pipefish have male "pregnancy," where young develop in the male's brood pouch, and the father provides all of the parental care. These species exhibit the reverse of the pattern observed in rough-skinned newts: males almost always have mates, whereas many females fail to obtain a mate. In some cases, this reversal of investment results in sexual dimorphism that is the converse of what we typically see. An example of this is observed in a group of shorebirds called phalaropes (Figure 3.18). Like seahorses, male phalaropes provide all of the parental care, including incubating eggs. Females fight each other for access to males, and males choose among females with which to mate. As a result, females are larger and more colorful than males.

Frank Fichtmueller/Shutterstock.com

Figure 3.18. Two phalaropes (*Phalaropus lobatus*) mating. Note that the male, on the top with wings spread, is less colorful than the female.

3.6: Interaction between different types of selection

As we observed in the previous section, there are a variety of reasons that a trait might be favored. Traits can affect fitness by increasing or decreasing an individual's ability to avoid predation, ability to obtain sufficient food, number of offspring they produce, or mating success, just to name a few. We also observed that a trait could have multiple effects on fitness-related aspects of an organism: the peacock's tail both increases mating success and decreases ability to avoid predation. This is a case where different types of selection are acting on the same trait. In this case, we would conclude that the benefits of the tail for mating success greatly outweigh the cost to avoiding predators. In this section, we will examine another case where sexual selection and another type of selection interact, and how the environmental context determines the outcome. We will examine this through the lens of research done on wild guppies (*Poecilia reticulata*).

3.6.1. Guppy coloration and biology

The guppies studied here (Figure 3.19) are native to South America, although they have been introduced around the world by humans. Males are brightly colored, whereas females are not. Males engage in courtship displays to entice females to mate; females will mate with multiple males. In the following activity, we will examine whether male coloration affects their mating success, based on the following two experiments by Houde (1987).

Napat/Shutterstock.com

Figure 3.19. A male guppy (*Poecilia reticulata*) showing the orange spots that are measured in the studies of Endler (1980) and Houde (1987).

In the first experiment, Houde chose a number of males and classified them in four groups (A–D) according to the amount of orange coloration they exhibited, with A being the most orange and D being the least orange. A female was exposed to a male of each category in random order and the length of time from when the male started to display until mating was recorded, as well as how many times the male displayed before mating. The experiment was repeated 10 times, each time with different males and females.

In the second experiment, four males of different orange-spottedness were placed with 10 unmated females. Houde recorded the number of displays for each male that evoked a response from a female during a 10-minute period each day for two days. She replicated the experiment with different fish eight times.

Table 3.3. Results of Houde's (1987) first experiment. Letters indicate category of male based on coloration (A = most orange; D = least orange). For each of the 10 trials, the numbers indicate the rank of each male in terms of time to mating and number of displays until mating (1 = fastest/ fewest displays). Dashes indicate that a male did not mate during that trial. A rank of 1.5 means that two males tied for the fewest displays.

Trial number	Time to mating				Displays to mating			
	A	B	C	D	A	B	C	D
1	1	2	3	4	1	3	2	4
2	1	2	3	4	1	2	3	4
3	1	-	2	3	1.5	-	1.5	3
4	1	-	-	-	1	-	-	-
5	1	-	2	3	1	-	2	3
6	2	3	1	-	2	3	1	-
7	-	1	-	-	-	1	-	-
8	1	2	-	-	1	2	-	-
9	1	2	-	-	1.5	1.5	-	-
10	1	2	-	-	1	2	-	-

Activity

1. *State the hypothesis and the null hypothesis being tested.*
2. *What do we predict the relationship should be between coloration and the time until mating, based on the alternate hypothesis? What should be the relationship between coloration and the number of displays until mating, based on the same?*
3. *What do we predict the relationships in #2 should be if the null hypothesis is true?*
4. *Table 3.3 shows the results of Houde's first experiment. Note that for each trial she ranks the different males by who was fastest to mating and by who had the fewest displays until mating. Given our predictions above, what do we expect these tables to look like for both the alternate hypothesis and the null hypothesis?*
5. *Note also that some males in the first experiment never mated. Which males do we expect to have failed to mate more often, based on our alternate and null hypotheses?*
6. *How do the results compare to our predictions? Which hypothesis is supported, and which should we reject?*
7. *Figure 3.20 illustrates the results of the second experiment as a scatter plot of the relationship between the amount of orange on a male (measured as the percent of the area of the body covered by orange) versus the number of responses a male evoked divided by the number of displays he made. What should this graph look like based on our alternate hypothesis? What should it look like based on the null hypothesis?*

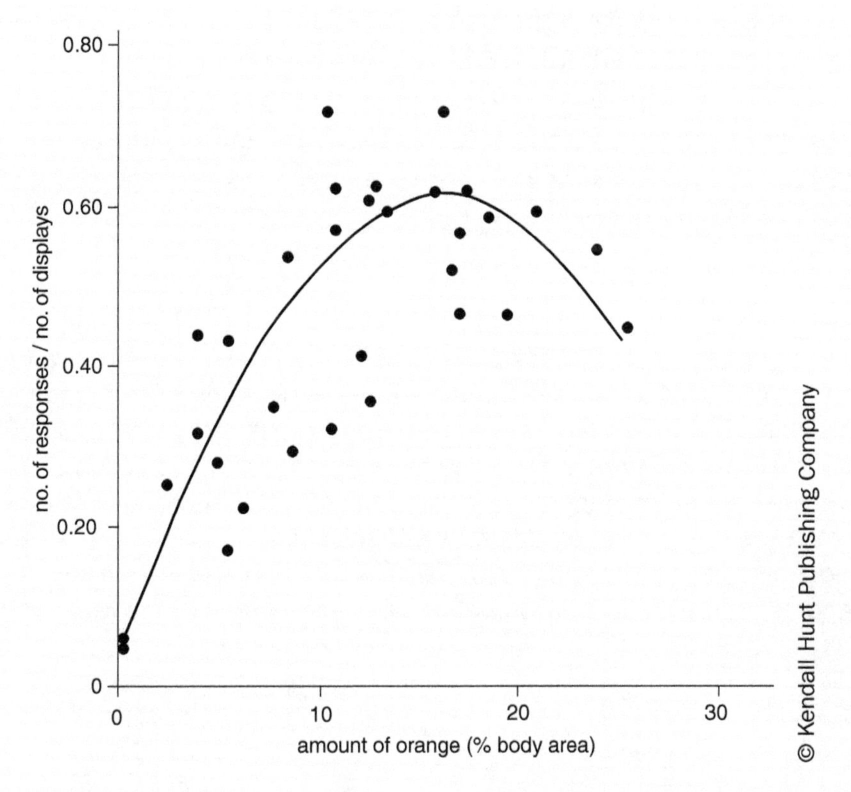

Figure 3.20. Scatter plot of amount of orange on a male (calculated as the percentage of its body covered by orange spots) versus the number of responses to a male by females divided by the number of displays made by the male. The curve is a quadratic curve of best fit to the data. (Modified from Houde, 1987: Fig. 3.)

8. *How do the results illustrated in Figure 3.20 compare to our predictions? Which hypothesis is supported, and which should we reject? How does this compare to our conclusions based on the first experiment?*
9. *Notice that Houde fit a curve to her data. If this curve is a good description of the relationship depicted in the data, what does this suggest about the effect on mating success of having more than about 20% of the male's body covered in orange? How might you explain this?*

Houde's experiments demonstrate that orange coloration is important for male mating success. But what effect does coloration have on predation, and how does this effect the evolution of male coloration? To examine this, we will look at the experiments of Endler (1980). Endler ran two experiments, one in a greenhouse and one in the field. In the greenhouse experiment, he set up 10 ponds with a number of guppies. After allowing the guppies to establish a population over several weeks, he then added one of two predators to some of the ponds. One was a large cichlid fish, *Crenicichla alta*, and the other was a smaller cyprinid, *Rivulus hartii* (now placed in the genus *Anablepsoides*). *Crenicichla* is considered to be a more effective predator of guppies than is *Rivulus*. Some ponds had no predators added. Endler took two censuses of the ponds, at 5 months and 14 months after the predators were added and measured the number and size of the orange spots on the males.

In the field experiment, guppies were transplanted from a pool that had *Crenicichla* present to a pool that had no guppies and no *Crenicichla* but that was inhabited by *Rivulus*. The two pools were sampled two years later and the male color patterns were recorded.

Activity

Figure 3.21 shows the results of Endler's greenhouse experiment. The x-axis represents time in months, with an additional legend indicating when the populations were founded (F), when the predators were added and the experiment started (S), and the two census points (I and II). The y-axis indicates the number of orange spots per male fish. The first two points on the graph summarize the number of spots for all ponds, and for the last two census points there are separate points for the three treatments: no predators (K), Crenicichlia *added (C), and* Rivulus *added (R). Note that the dots indicate the average number of spots and the bars indicate two standard errors above and below the mean (which captures more than 95% of the range of variation).*

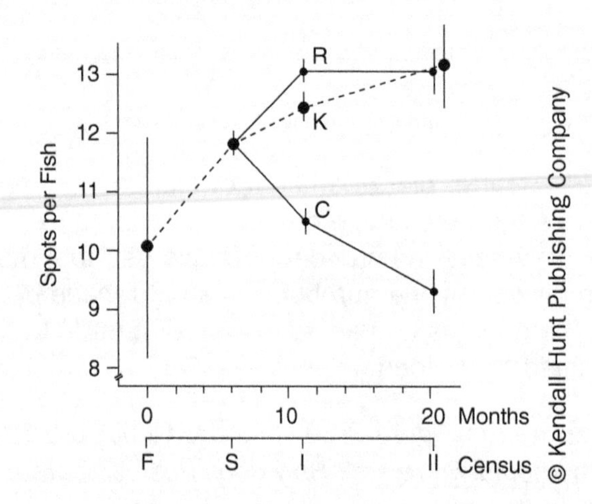

Figure 3.21. Graph of results of Endler's (1980) greenhouse experiment. Dots indicate mean number of spots per male fish, and error bars indicate two standard errors. In "Census" bar: F = time of foundation of guppy population; S = start of experiment, when predators are introduced into population; I and II = points when Endler took a census of the population and recorded number of orange spots on males. Data for F and S taken from all ponds; data for I and II for individual treatments, indicated as follows: K = no predators; R = *Rivulus* present; C = *Crenicichla* present. (Modified from Endler, 1980: Fig. 1.)

1. *How did the total population change from the foundation (F) until the start of the experiment (S), both in terms of the average number of spots and the range of variation in spots? Does this make sense given what we observed in Houde's experiments? Why or why not?*
2. *Which treatment, if any, in the greenhouse experiment is the control?*
3. *If selection due to predation is stronger than sexual selection on spots, what do we predict we should see in the graph? What should we see if sexual selection is stronger than selection due to predation?*
4. *Based on what you predicted above, what do you conclude regarding the strength of selection due to predation by* Crenicichlia *versus that of sexual selection? How about for the strength of selection due to predation by* Rivulus*?*

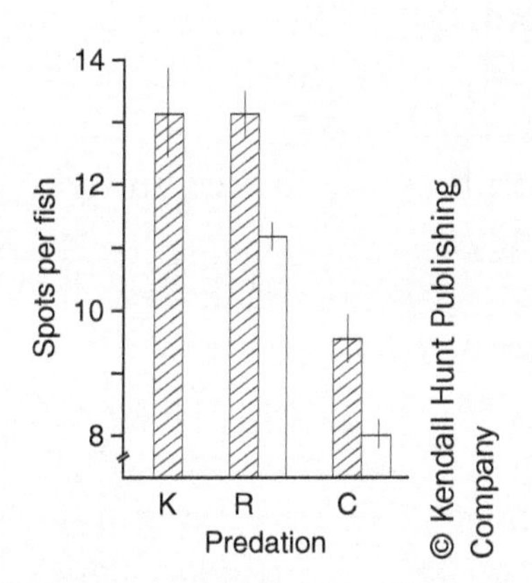

Figure 3.22. Graph of spots per male guppy at the end of Census II in Endler's (1980) greenhouse experiment (hatched bars) and in the field experiments (solid bars). Bar height indicates mean spots per fish, error bars indicate two standard errors. K = no predators; R = *Rivulus* present; C = *Crenicichla* present. (Modified from Endler, 1980: Fig. 2.)

Figure 3.22 summarizes the data on spots per fish for each of the three greenhouse treatments at Census II, as well as for the two field treatments.

1. *Does this graph support your conclusions for the greenhouse experiment from Figure 3.21?*
2. *Which treatment, if any, is the control in the field experiment?*
3. *Based on the results of the greenhouse experiment, how would you predict the two field experiment treatments would compare at the end of the experiment in terms of number of spots?*
4. *Do the results match your predictions?*
5. *If you were studying guppies in the wild, under what conditions would you expect to find that the males have relatively few spots?*

Endler's experiments demonstrate how different types of selection interact based on the same trait. When predation is strong, males with fewer spots might be favored. However, even in the presence of a predator, sexual selection in favor of spots might be strong enough to outweigh the detrimental effect on avoiding predation.

Literature Cited

Barrette, C., and D. Vandal. 1990. "Sparring, Relative Antler Size, and Assessment in male Caribou." *Behavioral Ecology and Sociobiology* 26:383–87.

Carlson, T. 1913. "Über Geschwindigkeit und Grösse der Hefevermehrung in Würze." *Biochemische Zeitschrift* 57:313–34.

Coltman, D W., P. O'Donoghue, J T. Jorgenson, J T. Hogg, C. Strobeck, and M. Festa-Bianchet. 2003. "Undesirable Evolutionary Consequences of Trophy Hunting." *Nature* 426:655–58.

Endler, J C. 1980. "Natural Selection on Color Patterns in *Poecilia reticulata*." *Evolution* 34:76–91

Houde, A E. 1987. "Mate Choice Based Upon Naturally Occurring Color-Pattern Variation in a Guppy Population". *Evolution* 41:1–10.

Jones, A., J. Arguello, and S. Arnold. 2002. "Validation of Bateman's Principles: A Genetic study of Sexual Selection and Mating Patterns in the Rough-Skinned Newt." *Proceedings of the Royal Society of London B* 269:2533–39.

Kettlewell, H B D. 1955. "Selection Experiments on Industrial Melanism in the Lepidoptera." *Heredity* 9:323–42.

Møller, A P. 1988. "Female Choice Selects for Male Sexual Tail Ornaments in the Monogamous Swallow." *Nature* 322:640–42.

Møller, A P. 1990. "Effects of an Haematophagous Mite on the barn Swallow (*Hirundo rustica*): A test of the Hamilton and Zuk Hypothesis." *Evolution* 44:771–84.

Peek, F W. 1972. "An Experimental study of the Territorial Function of Vocal and Visual Display in the male red-Winged Blackbird (*Agelaius phoeniceus*)." *Animal Behaviour* 20:112–18.

Smith, J., and A. Dhondt. 1980. "Experimental Confirmation of Heritable Morphological Variation in a Natural Population of Song Sparrows." *Evolution* 34:1155–60.

Further Reading

Darwin, C R. 1871. *The Descent of Man and Selection in Relation to Sex*, 475. London: John Murray.

Gross, M R. 1985. "Disruptive Selection for Alternative Life Histories in Salmon." *Nature* 313:47–48.

Scheffer, V B. 1951. "The Rise and Fall of a Reindeer Herd." *The Scientific Monthly* 73:356–362.

Weiner, J. 1995. *The Beak of the Finch: A Story of Evolution in Our Time*, 332. New York: Vintage Books.

CHAPTER 4

Testing Evolution at the Genetic Level

Evolution is often described in its most general form as "change over time." This definition might seem adequate, but it is insufficient if we want to know what we would observe if evolution was occurring. What is actually changing, and how would we know that it has changed?

To answer the first question, we need to clarify what constitutes *evolutionary* change over time. Individuals, for instance, change over time, as developing embryos or in the sense that most multicellular organisms increase in size from when they are fertilized eggs until they reach some adult size, or until they die; as humans, we are all aware of how much we have grown since we were children. This kind of change is what we usually call development or **ontogeny**, the change an individual experiences over its lifetime. Individuals also change in response to their environments and outside stimuli. Plants grow toward sunlight. Some mammals grow more fur in response to cold, or even change the color of their fur according to the season. Chameleons change the color of their skin to match their surroundings. In humans, exposure to the sun produces sun tans, working out enlarges muscles, and eating too much fat leads to obesity and other health issues. This type of change is not what we mean by evolution. It does not lead to new types of organisms or explain the diversity of life.

Darwin is credited with introducing a new way of thinking about organisms: population thinking. In other words, Darwin's ideas promoted thinking about populations, groups of individual organisms of the same species living in the same region. His work promoted thinking about how populations consist of individuals that differ from one another, and that populations could change over time not because individuals change, but because the numbers of different kinds of individuals in the population change. Polar bears became white not because brown-furred individuals changed their fur to white. Instead, some ancestral population consisted of individual bears that had different shades of fur color, including brown and white. Over time, the population came to consist of more and more white bears and fewer and fewer brown bears, until at some point the population consisted only of white bears. Thus, evolution concerns change in populations—not individuals—over time. The discussion of natural selection in Chapter 3 depends heavily on population thinking, from thinking about how populations grow to how heritable variation among individuals in a population will affect the next generation.

4.1 Phenotypic evolution—the case of Darwin's finches

So, what about the population is changing? The change could be in some observable trait—part of what we call the **phenotype**—that varies among individuals in the population. The trait could, for instance, be the depth of the beak of a bird, or the length of the leg of a lizard, or the color of the flowers of a plant. Traits do not have to be static bits of anatomy. They could be aspects of physiology, such as chemical pathways for obtaining energy or performing photosynthesis; the molecules that determine blood types; or how an organism obtains enough oxygen for its body to function. They could be behavioral traits, like songs or calls, or innate responses to a stimulus. What is important about these traits for our purposes is that they show variation—that individuals can differ from each other in terms of these traits. When compared with another, an individual could have a longer or shorter leg length, a different blood type, a song that sounds different, and so on.

One of the best examples of observed evolution of phenotypic traits comes from the studies of Galápagos finches by Rosemary and Peter Grant and their students and colleagues. Galápagos finches (sometimes called Darwin's finches, although they did not figure much in Darwin's work; Figure 4.1) include 14 species that diversified from an ancestor that reached the islands from mainland South America. At first glance, all of the species just look like a bunch of small, drab birds, which is not surprising given that the diversification of the birds is presumably no older than the islands, about five million years. A closer inspection, however, reveals that each species is specialized anatomically or behaviorally for a unique lifestyle. In particular, the shapes of the beaks of different species are related to differences in the birds' diets. Thus, the Galápagos finches are considered to be an example of a relatively recent **adaptive radiation**. Adaptive radiations are essentially when an ancestral species gives rise to many new species that are specialized for different ways of life.

Wilfred Marissen/Shutterstock.com

Figure 4.1. Medium ground finch (*Geospiza fortis*) from the Galápagos Islands.

The Grants and their colleagues have been studying finches on the Galápagos island Daphne Major, and their studies have focused on one species, *Geospiza fortis*, commonly known as the medium ground finch. The research team caught and banded every bird on the island in the first year of the study and have banded all the hatchlings produced ever since. As a result, they have been able to document how this population has changed over time. They have also carefully studied the plants, and particularly their seeds, on the island in an effort to examine how changes in the availability of different seeds affected the population.

In 1977, a drought befell the islands, and this had a dramatic effect on the animals and plants. The finch population was drastically reduced. Among the plants, those that produced large, hard seeds tended to fare better than those that produced small seeds. The finches are seed-eaters, and cracking bigger, harder seeds is difficult without a larger, deeper beak. With this in mind, the researchers investigated the question of how the finch population changed in response to the drought.

4.1.1 Constructing hypotheses

Like any other scientific investigation, we can put this question into the context of the scientific method. In the context of this chapter, our real question is: Did the finch population evolve? Thus, we could articulate the following hypothesis.

H_A: The finch population evolved in response to the drought.

Based on what we learned in Chapter 3, we might further hypothesize that finches better able to eat large, hard seeds would be more likely to survive. But, for now, let's stick with this basic hypothesis. We can now also come up with a null hypothesis.

H_0: The finch population did not evolve in response to the drought.

4.1.2 Making predictions

To test these hypotheses, the researchers wanted to compare measurements taken before and after the drought. Among the measurements they took was the depth of each finch's beak. Figure 4.2 shows the distribution of beak depths in the population in 1976, before the drought. Note the shape of the graph; this "bell curve" is also called a **normal distribution** and is how many trait measurements are distributed in a population—most individuals have values for their traits that fall near the center of the range of possible values, and fewer and fewer individuals have each trait value as you approach either extreme. From these data we can also calculate the **mean**, or average value, for beak depth in the population; we reviewed this calculation in Chapter 1. One way we can determine if the population has changed in terms of beak depth is to look for a change in the mean value for beak depth.

Activity

Based on the alternate and null hypotheses articulated above, predict what the population should look like after the drought. How would you expect the population's mean beak depth before the drought to compare with its mean beak depth after the drought? How would the graph change (or not change) based on your predictions? Draw graphs representing your predictions.

If our null hypothesis is true and the population is not evolving, then we expect the population after the drought to have a graph of the distribution of individual beak depths that looks very much like the 1976 normal distribution graph. Of course, we do not expect these two graphs to be exactly the same; we already know, for instance, that there will be many fewer birds overall after the drought. But we can still predict, based on the null hypothesis, that the shape of the normal distribution graph will be similar, and, more importantly, the mean value for beak depth should not have changed substantially from the 1976 mean beak depth value.

If our alternate hypothesis is true and the population is evolving, we can make some different predictions. Most importantly, the mean beak depth of the population will be different before and after the drought. Given that we have not specified in our hypothesis how the population will evolve, we can predict that the mean will be either lower or higher. Again, given the connection between seed size and beak shape we made in the last chapter, you would predict a change in one direction and not the other. But for now, we're starting with the simple prediction that beak depth changed as a result of the drought.

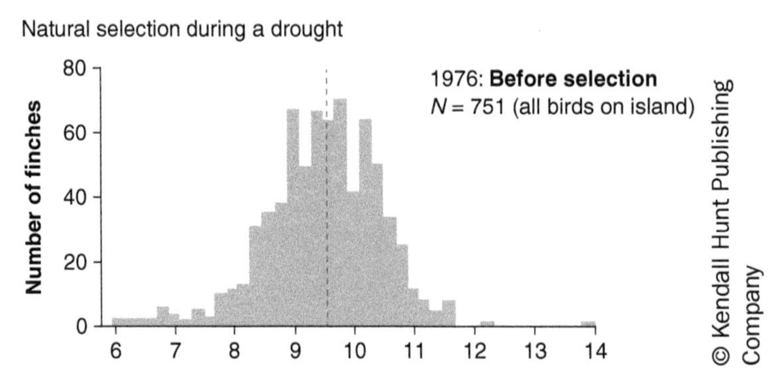

Figure 4.2. A graph of the distribution of beak depths for medium ground finches on Daphne Major in 1976. Red dashed line indicates the mean for the population.

4.1.3 Comparing results and predictions

Figures 4.2 and 4.3 represent the actual measurements of the finch population in 1976 and 1978, respectively. The mean beak depth of the population was significantly higher in 1978, which matches the predictions for our alternate hypothesis (and which, perhaps more importantly, does not match the predictions of our null hypothesis).

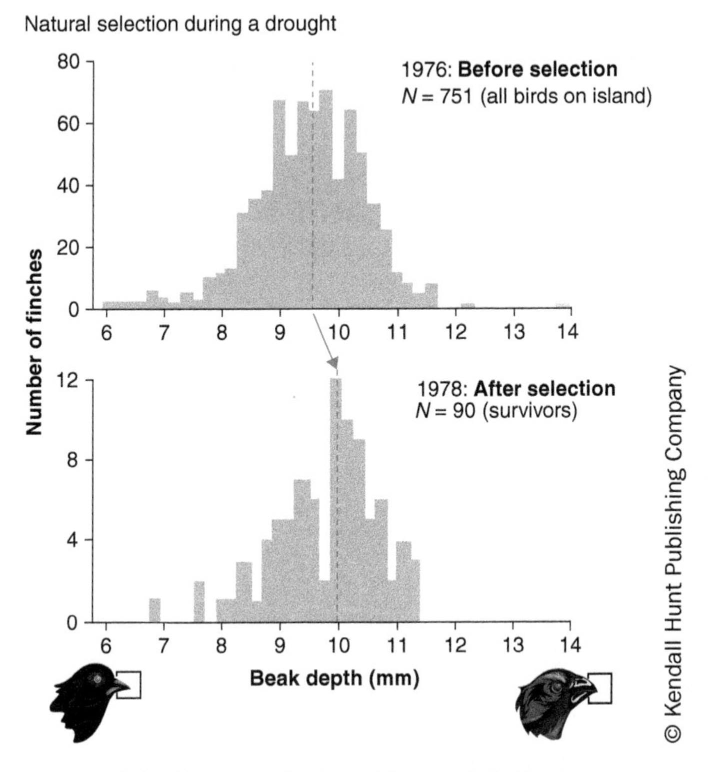

Figure 4.3. The graph from Figure 4.2 (top) combined with the same graph for 1978, showing the change in the mean beak depth for the population.

4.2 Natural selection and population genetics

In Chapter 3, we focused almost entirely on natural selection as reflected in phenotypes. The reason this book begins with this is because (1) natural selection is familiar to some extent to almost all students, and (2) it is fairly easy to understand the relationship between phenotypic traits and selection. While students might still come to this book with misconceptions about natural selection, it is easier for them to correct these misconceptions with straightforward examples like Kettlewell's experiments on peppered moths. The role of heritability in evolution by natural selection implies that selection has implications not only for changes in phenotype in a population, but also for changes in the genes that underly those traits. Now that we've examined selection in terms of phenotypes, we will now add another layer of complexity by examining the implications for the evolution of genes.

A closer look: Mendelian genetics

This chapter uses terminology based on Mendelian genetics, named for Gregor Mendel, an Augustinian friar living in what is now the Czech Republic. He was Darwin's contemporary, though his work was not appreciated until the twentieth century. Mendel gave us the first clear evidence that inheritance consisted of discrete elements, what we now think of as genes. Some of his ideas have been greatly modified in light of what we now know from molecular genetics and DNA, but his work paved the way for modern genetics.

Before Mendel, inheritance of traits was rather mysterious, and it was an issue that posed numerous difficulties for Darwin. The prevailing theory of inheritance was called blending, where the inherited material became mixed in the offspring in a manner where the parental trait could not be recovered. You can visualize this notion of inheritance as something like blending paint: red and blue paint blend to make purple, but you can never get red and blue back again. This idea had a number of problems, though, including the observed fact that some parental traits might reappear in "unblended" form after a few generations.

Mendel introduced and provided evidence for the idea that offspring inherited discrete units and discrete forms of those units, what we now call genes and alleles, from their parents. The term **gene** here refers to the basis of inheritance for a specific trait. We now know that genes ultimately correspond to sequences of DNA, and that genes often don't operate in such a simple fashion as one gene affecting one trait, but for our purposes this simple definition will suffice. A gene might have different forms, referred to as **alleles**. So, there could be a gene for the color of a flower, and that gene has two alleles, one that codes for yellow flowers and one that codes for red flowers. In sexually reproducing organisms, the offspring receives one allele from each parent. The combination of alleles that an individual possesses is called its **genotype**. Thus, the offspring could inherit the same allele from each parent (making its genotype **homozygous**) or a different allele from each parent (giving a **heterozygous** genotype). Mendel also demonstrated that some alleles exhibit **dominance** over other alleles (which are termed **recessive**), such that the phenotype associated with dominant alleles is always observed when the dominant allele is present in an individual. Thus, recessive phenotypes are only observed when the individual inherits a recessive allele from each parent. Note that not all genes exhibit dominance, and in some cases heterozygous individuals have a distinct phenotype from homozygous individuals. Note also that "dominant" has nothing to do with having a higher fitness, nor will the dominant allele necessarily be more common in the population; dominance is strictly about which allele is expressed in heterozygotes.

Mendel developed several "laws," the most important of which are segregation and independent assortment. Everyone recognized that an individual inherits from both parents, but Mendel recognized that the copies of genes inherited from the father and mother remain distinct, and that an individual will pass on only one those two inherited copies of a gene to its own individual offspring, without any "blending." This is his law of segregation. His law of independent assortment set out that genes governing different traits segregated independent of one another; in other words, just because you passed on to your offspring a copy of one gene from your father does not mean that all of the genes you pass on to that offspring will come from your father. We now know that this is only partly true, as some genes are "linked" on the same chromosome.

We used Kettlewell's experiments on peppered moths to illustrate natural selection on a phenotypic trait, and these kinds of experiments have been replicated in other species, and they have been extended to the genetic basis of the traits in question. One case that has clear parallels with that of the peppered moths involves rock pocket mice of the desert Southwest of the United States (*Chaetodipus intermedius*). Like the moths, rock pocket mice have light and dark morphs. Their desert range is dominated by light, sandy substrates, but a large part of the range of these mice is an expanse of ancient lava turned to dark basalt. The light morphs blend in with the sandy substrate but are conspicuous on the basalt, whereas the dark morphs are cryptic on the basalt and conspicuous on the sand. The color of the mice is governed by a gene called MC1R. The gene has two alleles, dark (D) and light (d), where D is dominant to d.

Activity

1. *Come up with an alternate hypothesis regarding how the color morphs will affect survival of individual mice in different parts of the range.*
2. *State the null hypothesis for the alternate hypothesis you proposed for no. 1.*

Analogous to what we saw with the moths, we would predict that light-colored mice would be favored in sandy areas, whereas dark mice would be favored on the basalt. Researchers studying these mice examined the phenotypes at several sites in the range of the mice, including some locations in the sandy desert and others on the ancient basalt flows. They also tested the genes of the sampled mice to determine what genotypes they had. They summarize these data in Figure 4.4 as **frequencies**, basically the fraction of the population with that phenotype, genotype, or allele.

Activity

1. *Based on the alternate hypothesis you generated in #1 of the previous activity, what do you predict the frequencies of the light and dark morphs should look like for sandy and for basalt locations?*
2. *Based on the null hypothesis you stated in #2 of the previous activity, what do you predict the frequencies of the light and dark morphs should look like for sandy and for basalt locations?*
3. *Based on the alternate hypothesis you generated in #1 of the previous activity, what do you predict the frequencies of D and d alleles should look like for sandy and for basalt locations?*
4. *Based on the null hypothesis you generated in #1 of the previous activity, what do you predict the frequencies of D and d alleles should look like for sandy and for basalt locations?*
5. *Compare your predictions to the results depicted in Figure 4.4. Do the results support the alternate hypothesis or null hypothesis?*
6. *Notice that the "Mid" population is 100% dark phenotypes, but the frequency of the D allele is not 1.0. Why do think this is the case?*

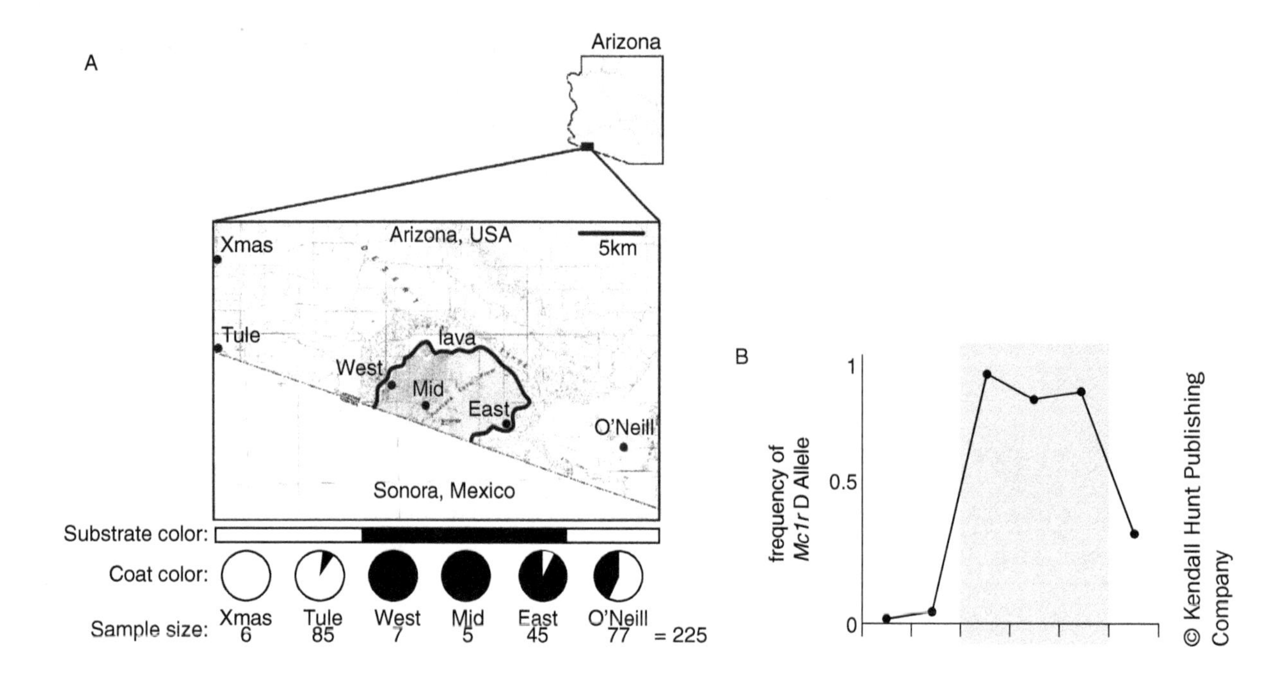

Figure 4.4. (A) Pie charts showing the frequency of light (white) and dark (black) morphs of rock pocket mice from different localities in Arizona. The location of the ancient basalt flow with dark substrates is outlined in the middle of the map. (B) Frequency of the dark (*D*) allele at the same localities from (A). Shaded area indicates the localities with dark substrate.

The results of the study match the predictions we would make based on a hypothesis of coat color being selected for based on the environment. Dark phenotypes dominate the basalt flows, and light morphs dominate the sandy areas. Similarly, the frequency of the D allele is very high in the basalt and very low in the sandy regions.

4.3 Evolution as genetic change in a population

Instead of changes in phenotypic traits, we could characterize a population in terms of changes in genes. Genes have two important properties. First, they determine, at least to some extent, what traits an individual has. For instance, there is a gene that governs the ABO blood groups that characterize our blood types, and that gene has three alleles (A, B, and O). Our individual blood types are determined by the alleles that we possess for that particular gene. Each individual has two copies of the blood group gene, and therefore the **gene pool** for the population includes a number of copies of this gene equal to twice the number of individuals.

Second, genes are passed from parent to offspring, so the offspring's phenotype is determined by which alleles it inherited from the parents. Since we inherit genes from both of our parents, we end up with two copies of our blood type genes, and the alleles that we inherit determine our blood type. The different pairings of alleles that an individual can possibly inherit from its parents are called **genotypes**. The AA and AO genotypes are blood type (i.e., phenotype) A, genotypes BB and BO are type B, genotype AB is type AB, and genotype OO is type O. Note that, although all three alleles might be present in the population's gene pool, a given individual cannot have more than two alleles in its genotype for this gene.

Since genes determine many of the traits in which we are interested, we can measure changes in a population's genes instead of changes in phenotypes. One way we can measure genetic change in a population is to look for changes in **frequencies** of alleles. A frequency is simply the fraction of the population exhibiting whatever we are measuring. Thus, we can calculate frequencies for different phenotypes, alleles, or genotypes. Note that phenotype and genotype frequencies would be calculated by dividing the number of individuals with a given phenotype or genotype by the population size (i.e., the number of individuals in the population), whereas the frequency of alleles would be calculated as the number of copies of that allele present in the population divided by the total number of copies of that gene in the gene pool (or twice the population size). Note that the sum of the frequencies of whatever we are calculating must always be 1.0.

Just to make it clearer, let's put these frequencies in the form of formulas. Consider a population of organisms and one of their genes. The gene has two alleles; *A* and *B*. The three genotypes are *AA*, *AB*, and *BB*; each genotype has a different phenotype. The number of individuals in the population is equal to *N*. Given this, here are some examples of how we can calculate different frequencies.

$$\textit{Frequency of genotype AB} = (\#\ \textit{of AB individuals})/N$$

Note that, in this case, the frequency of the *AB* genotype is also the frequency of its associated phenotype.

$$\textit{Frequency of allele A} = (\#\ \textit{of copies of A in the population})/2N$$

If we know the number of individuals with each genotype, we can calculate the number of *A* alleles and calculate the frequency of *A* by adding twice the number of *AA* individuals to the number of *AB* individuals. This is because each *AA* individual carries two copies of the *A* allele, whereas each *AB* individual has only one copy. Thus

$$\textit{Frequency of allele A} = \left[2(AA)+(AB)\right]/2N$$

Activity

1. *Using the example given above, determine the formulas for the frequencies of the AB and BB genotypes and the B allele.*

Wild chili pepper plants (Genus Capsicum*) exhibit heritable variation in the "heat" of the peppers, which is related to the amount of the chemical capsaicin they produce. Let's assume that the levels of capsaicin in these peppers can be classified into three phenotypes: "mild," "medium," and "hot." Let's also assume that the level of heat is governed by a single gene locus with two alleles ("hot" and "mild"), with heterozygous individuals having the intermediate phenotype ("medium").*

Two populations of wild chili peppers were studied and provided the data summarized in Table 4.1.

2. *What are the frequencies of the three phenotypes (mild, medium, and hot)?*
3. *Calculate the frequencies of the mild and hot alleles and of the mild, medium, and hot genotypes.*

Table 4.1. Data on two hypothetical populations of chili pepper plants that include mild, medium, and hot individuals due to a hypothetical gene with two alleles, "mild" and "hot."

	Total population	Mild	Medium
Population A	1276	204	615
Population B	4563	930	1990

Hopefully you noticed that you could calculate the number of "hot" individuals by subtracting the numbers of "mild" and "medium" individuals from the total population.

4.4 The Hardy–Weinberg Equilibrium as our null hypothesis

Evolution is commonly described as *changes in allele frequencies over time*, so we can look for changes in allele frequencies as evidence that a population is evolving. If we can follow a population over time, we can then ask whether allele frequencies change from what we predict if there was no evolution. But what should we predict if there is no evolution? That, after all, is our null hypothesis. It might seem intuitive that the null hypothesis predicts that frequencies should stay the same. In fact, we can demonstrate this prediction in a mathematical model. The model tells us what we expected: that absence of evolution means things do not change. But it also allows us to do something rather unexpected, namely to test whether a population is evolving without having "before" and "after" pictures of its status. The model that gives us the predictions for our null hypothesis is called the **Hardy–Weinberg Equilibrium (HWE)**.

A closer look: Hardy–Weinberg, Punnett squares, and Mendel

The British mathematician G. H. Hardy and the German physician Wilhelm Weinberg independently published what we now know as the Hardy–Weinberg principle in 1908. The problem was introduced to Hardy by the British geneticist Reginald Punnett, creator of a tool used in genetics called a **Punnett square**. Punnett squares are actually useful for illustrating the derivation of the Hardy–Weinberg principle. Punnett squares illustrate the expected frequencies of genotypes of offspring produced by two parents based on Mendelian genetics.

We can represent how Mendelian inheritance works using Punnett squares. For our purposes, we'll be looking at the simplest case, considering only a single gene with two alleles.

A Punnett square is essentially a table with the possible parental gametes—the sex cells, sperm and eggs—indicated on the rows and columns, and the cells are filled with the resulting genotypes by combining the gametes. Each parent can actually produce many sperm or eggs, but each sperm or egg will carry only one allele for a particular gene. If a parent is homozygous for a particular gene, all of his or her gametes will have the same allele for that gene. If a parent is heterozygous, then half of the gametes will have one allele and the other half will have the other allele.

Table 4.2 gives an example of a Punnett square using two heterozygous parents, where the dominant allele is *A* and the recessive is *a*. The gametes are represented by the boxes with dark borders, with the red letters representing the gametes of one parent and the blue letters representing the gametes of the other parent.

Table 4.2. An example of a Punnett square for two parents that are heterozygous for a gene with alleles *A* and *a*.

	A	**a**
A	*AA*	*Aa*
a	*aA*	*aa*

Note that "*Aa*" and "*aA*" are the same genotype. Note also that what this shows us is the *possible* genotypes that can be produced by this mating, and the *expected* ratios in which they should be produced. Thus, we expect twice as many heterozygous offspring as we do offspring with either of the homozygous genotypes. In terms of phenotypes, since *A* is dominant, we expect three out of four offspring to have the dominant phenotype.

Punnett squares are very helpful for determining what should come of a mating of two individuals, but we are interested in how populations should behave. What proportions of alleles, genotypes, and phenotypes should we expect in the next generation of a population? One helpful thing to note is that the columns and rows of the Punnett square in Table 4.2 essentially describe a gene pool with allele frequencies of 0.5 for both alleles. If we instead redrew the Punnett square where each parent had five columns or rows for *A* and five columns or rows for *a*, we would still come to the same conclusions regarding the expected proportions of offspring with different genotypes or phenotypes. Because an individual can only carry one (if homozygous) or two (if heterozygous) alleles, then the frequency of the two alleles in that individual's gametes will be either 1.0 and 0.0 (for homozygous) or 0.5 and 0.5 (for heterozygous). We can think of a population in the same way we think of individuals, except that a population could have a wide range of possible allele frequencies.

Let's extend the Punnett square to a population, illustrated in Table 4.3. Imagine a population where the frequency of allele *A* is 0.7 and that of allele *a* is 0.3. Instead of two parents each with equal numbers of gametes carrying each allele, think of the population as two parents where 7 out of 10 gametes carry *A* and 3 out of 10 carry *a*.

Table 4.3. A Punnett square representing a population where the frequency of allele *A* is 0.7 and that of allele *a* is 0.3. The colors correspond to parts of the table with the same genotype.

	A	A	A	A	A	A	A	a	a	a
A	AA	AA	AA	AA	AA	AA	AA	Aa	Aa	Aa
A	AA	AA	AA	AA	AA	AA	AA	Aa	Aa	Aa
A	AA	AA	AA	AA	AA	AA	AA	Aa	Aa	Aa
A	AA	AA	AA	AA	AA	AA	AA	Aa	Aa	Aa
A	AA	AA	AA	AA	AA	AA	AA	Aa	Aa	Aa
A	AA	AA	AA	AA	AA	AA	AA	Aa	Aa	Aa
A	AA	AA	AA	AA	AA	AA	AA	Aa	Aa	Aa
a	aA	aA	aA	aA	aA	aA	aA	aa	aa	aa
a	aA	aA	aA	aA	aA	aA	aA	aa	aa	aa
a	aA	aA	aA	aA	aA	aA	aA	aa	aa	aa

There are several things to note here. First, as indicated by the shading, you can see the relative proportions of the three genotypes. You can calculate the number of boxes for each genotype the same way you can for any array, by multiplying the number of rows by the number of columns. This gives you 7 × 7 = 49 *AA*, 3 × 3 = 9 *aa*, and (3 × 7) + (7 × 3) = 42 *Aa*. With 100 total boxes, it is easy to convert these numbers of boxes into frequencies by dividing by 100, giving you: *AA*: 0.49, *Aa*: 0.42, and *aa*: 0.09; you'll notice that these add up to 1. Notice also that the frequency of *AA* is equal to the initial frequency of *A* (0.7) squared, the frequency of *aa* is the initial frequency of *a* (0.3) squared, and the frequency of *Aa* is twice the product of the initial frequencies of *A* and *a* (or 2 × 0.7 × 0.3). Since not every population will have frequencies of 0.7 and 0.3, let's replace those numbers with variables, so we can apply this to any population for a gene with two alleles. If we use *p* to represent the frequency of *A* and *q* to represent the frequency of *a*, we can represent the genotype frequencies and their sum by the formula $p^2 + 2pq + q^2 = 1$.

Now we can also calculate the allele frequencies in our new generation. Out of the 100 individuals represented in the Punnett square, 49 carry two copies of *A* and 42 carry one copy, so the frequency of *A* in the new generation is:

$$p = [(2 \times 49) + 42] / 2(100) = (98 + 42) / 200 = 140 / 200 = 0.7$$

We can calculate the frequency of *a* in the next generation in the same manner, or by simply subtracting 0.7 from one, because the frequencies of the two alleles must add up to one; to put it as a formula:

$$p + q = 1$$

Either method tells us that the frequency of *a* is 0.3. The original allele frequencies were also 0.7 and 0.3, so the allele frequencies did not change. That means the Punnett square for the next generation will also look like the one above. Same for the fourth generation, same for the millionth generation. In other words, what we've just modelled is that allele and genotype frequencies should stay the same in perpetuity—unless something funny happens, like something that causes evolution.

Let's apply the HWE to illustrate what it tells us about the evolution of a population. Note that we're going to model one of the simplest situations: we're looking at a single gene with two alleles. It's possible to apply HWE to multiple genes, or to a gene with more than two alleles, or to both kinds of complications, but the simple model we are using here is good for illustrating our concepts. In this example, we have two variables that go into the HWE formulas: p represents the frequency of one allele and q represents the frequency of the other.

HWE is usually distilled into two formulas. The first is fairly easy to grasp, when you consider that when there are only two alleles for a gene, if some percentage of the gene pool has one allele, the remainder must have the other one. Mathematically

$$p+q=1$$

The derivation of the second formula is described in the box on population genetics above, and what we need to understand is that the terms in this formula correspond to the frequencies of the genotypes. Specifically p^2 is equal to the frequency of homozygous individuals with one allele (the one whose frequency is p), q^2 is equal to the frequency of homozygous individuals with the other allele (the one whose frequency is q), and $2pq$ is the frequency of heterozygous individuals. Since those frequencies cover all the individuals in the population, we get:

$$p^2+2pq+q^2=1$$

To illustrate how we use HWE, let's use the following hypothetical example from an earlier activity. Wild chili pepper plants (Genus *Capsicum*) vary in the "heat" of the peppers, which is related to the amount of the chemical capsaicin they produce. Let's assume that the levels of capsaicin in these peppers can be classified into three phenotypes: "mild," "medium," and "hot." Let's also assume that the level of heat is governed by a single gene with two alleles ("*H*" and "*M*"), with heterozygous individuals having the intermediate phenotype ("medium"). A hypothetical study of a population reported the numbers of individuals with each phenotype given in Table 4.4.

Table 4.4. Results of a hypothetical study of a hypothetical chili pepper plant population showing the number of individuals with each of three phenotypes.

Total population	Mild	Medium	Hot
1276	204	615	457

We can calculate the frequencies of the alleles and genotypes for the population using the formulas we used before. For instance, the frequency of the mild genotype would be equal to

$$MM \text{ frequency} = 204/1276 = 0.16$$

The frequency of the H allele would be

$$H \text{ frequency} = [(2\times 457)+615]/(2\times 1276) = 0.60$$

The frequencies for both alleles and all three genotypes are given in Table 4.5.

Table 4.5. Observed frequencies of alleles and genotypes for the hypothetical population from Table 4.4.

Frequency of				
Mild allele	**Hot allele**	**Mild genotype**	**Medium genotype**	**Hot genotype**
0.40	0.60	0.16	0.48	0.36

Note that these are the *actual* frequencies that we have calculated from our observations of the population. At this point, we have not used any of the HWE formulas to determine anything. Now that we know the actual frequencies of the alleles, we can see if this population matches the predictions of HWE. Recall that HWE predicts that allele and genotype frequencies should stay the same from generation to generation. Thus, if this population were in HWE, the allele frequencies that we calculated for this generation would be the allele frequencies of the previous generation as well. And if we plug those frequencies into the HWE formula, we should get the same genotype frequencies that we calculated above. So, if the frequency of H is p, and the frequency of M is q, and we plug our calculated allele frequencies into

$$p^2 + 2pq + q^2 = 1$$

we get

$$(0.6)^2 + 2(0.6)(0.4) + (0.4)^2 = 1$$

or

$$0.36 + 0.48 + 0.16 = 1$$

These three values—0.36, 0.48, and 0.16—are our expected values for the frequencies of the genotypes *HH*, *HM*, and *MM*, respectively. When we compare these values to the genotype frequencies we calculated, they look identical, which is what HWE would predict. In other words, we have demonstrated that this population is in HWE; it is therefore not evolving.

We can also use the genotype frequencies that we expect from HWE to determine how many individuals in the population should have each genotype. To do this, we simply multiply each genotype frequency by the population size. For instance, HWE's predicted number of hot individuals would be

$$0.36(1276) = 459.36$$

Notice that this number is not exactly the same as what we observed, but it is very close, close enough that we probably wouldn't change our minds about the population being in HWE. Table 4.6 gives the predicted number of individuals for all three genotypes.

Again, the numbers of individuals that we expected to get are not exactly the same but very close to the numbers that we actually observed. Thus, this population looks like it is in HWE.

Table 4.6. Predicted genotype frequencies and predicted and observed numbers of individuals for the hypothetical chili pepper plant population from Table 4.4.

Genotype	Predicted frequency	Predicted number of individuals	Actual number of individuals
HH	0.36	459.36	457
HM	0.48	612.48	615
MM	0.16	204.16	204

Activity

*Let's now apply HWE to an example based on a well-studied population of moths. Wing coloration in the scarlet tiger moth (*Callimorpha dominula*) is governed by a single gene with two alleles, which we'll call S and s. SS individuals are white-spotted. Individuals with ss have little spotting, while heterozygous individuals show an intermediate condition. (Since heterozygous individuals are different from both types of homozygous individuals, neither allele is dominant.)*

A sample of a population of these moths recovered the following numbers of each phenotype: white-spotted (SS) = 1469, intermediate (Ss) = 138, little spotting (ss) = 5.

1. *Calculate the **frequencies of each allele and each genotype** and record them in a table.*
2. *Use the allele frequencies you calculated above and the Hardy–Weinberg equations to calculate the **expected genotype frequencies assuming HWE**.*
3. *Use the genotype frequencies from #2 to calculate the **expected numbers of individuals for each genotype**.*
4. *How do the expected genotype frequencies that you just calculated compare to the observed genotype frequencies you calculated in #1? How do the expected numbers of individuals compare to what was actually observed?*

Hopefully, you noticed that what you calculated using the HWE formulas came very close to what was actually observed in the population. Of course, your numbers may have been a bit different, but, as in the chili pepper example, the differences were small enough that your calculations were essentially the same as what you observed. You might ask, though, "Just how different would my numbers need to be to decide that my calculations were not very close to what was observed?" You might also wonder, "If I plug the actual allele frequencies into the HWE equations, will I always get expected genotype frequencies that match the actual genotype frequencies?"

Activity

In this example, we have a single gene with two alleles, S and A, in a human population in Nigeria. We'll say more about this gene later, but for now we just need to be able to calculate some frequencies.

1. *In Table 4.7, fill in the observed allele frequencies for the population (Freq S and Freq A).*
2. *Then use the HWE formulas to fill in the expected genotype frequencies for SS, SA, and AA genotypes.*
3. *How do your expected values compare to the observed values?*

Table 4.7. Table of observed and expected numbers of individuals based on HWE, as well as actual allele frequencies, for a gene with two alleles, S and *A*, in a human population from Nigeria. Use the numbers provided to calculate the values for the blank cells.

		SS	SA	AA	Total	Frequency S (p)	Frequency A (q)
Nigerian population	**Observed**	29	2993	9365	12,387		
	Expected						

You probably noticed that your calculations of what you expected and what was actually observed were quite different. This raises three very important points.

1. First, just because you know p and q, it does not necessarily follow that you can discover the actual genotype frequencies using $p^2 + 2pq + q^2 = 1$. In HWE, a population with certain allele frequencies begets a population with the same allele frequencies, and so on. But not every population is in HWE.
2. Why isn't the population in HWE? In short, it's because the population is evolving. What is causing it to evolve is another question.
3. How do we know that the differences that we saw between observed and expected mean that the population is not in HWE? Why can't we treat them like the differences we saw in the moth example? That's what we will discuss next.

4.5 Is this population evolving?

Hardy-Weinberg provides us with our null hypothesis (i.e., "the population is not evolving"), but how do we determine if the differences we measure are actually significant enough to be called evolution? We seemed content to accept the differences we observed in the chili pepper and moth examples as just "noise" in the data, but the larger differences we observed for the Nigerian human population seemed to require some other explanation. How can we draw a line between "noise" and significant differences? The answer is that statistical tests allow us to test the likelihood that the differences between what we observe and what Hardy-Weinberg predicts are actually due to chance. If this likelihood is very low, then we can infer that the population is evolving. If this likelihood is not very low, then we cannot reject the null hypothesis that the population is not evolving. We can demonstrate how we can do this, using the cases we just discussed.

The statistical test we will use here is called the **chi-square test**. This test compares what we observe against what we expected to get on the basis of a given hypothesis. In this case, we're testing the hypothesis inherent in HWE, which is that a population is not evolving. To perform the test, we first calculate a statistic called chi-square, whose symbol is $\boldsymbol{X^2}$. (Note that X^2 is the symbol; it is composed of the Greek letter a X with a superscript "2," but it is not a variable that has been squared, so we won't be taking the square root of it.) The formula for calculating this statistic is:

$$X^2 = \sum \frac{(\text{observed} - \text{expected})^2}{\text{expected}}$$

Thus, what you do to calculate this statistic is first calculate what you expected to get, in our case the number of individuals for each genotype as predicted by HWE. Next, calculate the difference between the actual value observed (in our case, the actual number of individuals for each genotype) and the expected value. Square this difference and divide it by the expected value. Once you've done this for each category (for us, each genotype), add all of those calculations together (the Σ means "sum"). Thus, we will have three categories for which we will make these calculations: the observed and expected numbers of individuals for each of the three genotypes. **Note that we want to use *numbers of individuals*, not frequencies for these calculations**.

Activity

Go back to the calculations that you made for the activities in Section 4.3 concerning the moths and the Nigerian human population. Fill in the empty cells in Table 4.8 to calculate the chi-square statistic for each one.

How do the chi-square statistics of the two examples compare? What is driving the differences between them?

Table 4.8. Table for calculating the chi-square statistic for the data on scarlet tiger moths and humans in Nigeria. O, observed number of individuals; E, expected number of individuals. Fill in the table to calculate the statistic for each example.

	Genotype	Observed (# individuals)	Expected (# individuals)	O(E	(O-E)^2	(O-E)^2/E
Moths	**SS**					
	Ss					
	ss					
					Sum =	
Nigerian human population	**SS**					
	SA					
	AA					
					Sum =	

Once you've calculated a chi-square statistic, the next step is to figure out what it tells you. You probably noticed from comparing the two examples that the bigger difference between observed and expected in the Nigerian human population resulted in a much larger chi-square statistic than for the moths. The bigger the difference between observed and expected, the less likely that the difference is due to "noise" in the data. Thus, a larger chi-square value means a greater likelihood that the differences are significant and not "noise." But how big does it need to be to tell that the differences are significant?

The benefit of using the chi-square test is that it can be applied to a variety of cases where we are comparing predictions of a hypothesis with observed data. For any situation where we want to use the chi-square test, we need to calculate the chi-square statistic, and we need two other pieces of information. The first is the **degrees of freedom**, which is calculated as the **number of classes of data** minus the **number of independent variables**. For our examples, we had three classes of data (the three genotypes). The independent variables are the things we needed to know in order to come up with our predicted genotype frequencies from HWE. To accomplish this, we needed to know one allele frequency and the population size. We only needed to know one allele frequency, because if we knew p, we could determine q as $p-1$. That allowed us to calculate the genotype frequencies, and then we needed to multiply them by the population size to get the number of individuals. Thus, we had three classes and two independent variables, so our degrees of freedom are 3–2 = 1.

Besides the degrees of freedom, we need to know the ***p* value**, which is the lowest acceptable probability that the differences between observed and expected are due to chance alone. We will use a common standard for biological research of $p > 0.05$. In other words, if we determine that there is a greater than 5% probability that the differences between the observed and expected values are due solely to chance, then we can accept (or at least not reject) HWE as explaining what we observed in the population. In other words, we would have no evidence to support that the population was evolving.

If you look at a statistical table of chi-square critical values for the entry corresponding to one (1) degree of freedom (three genotypes minus two variables) and a p value of 0.05, you will find the critical value of $\chi^2 = 3.841$. Keep in mind that a small chi-square statistic should be more likely to indicate that chance is solely responsible for the differences between observed and expected, whereas a large value should be more likely to indicate that something other than chance has caused the differences. Thus, we can use the value of 3.841 as the boundary between these two explanations. If the calculated chi-square statistic (χ^2) is greater than this critical value, then we can say that chance is not likely to be responsible for the differences between the observed and expected values, which implies that evolution is occurring.

Activity

Compare the chi-square statistics you calculated in the previous activity to the critical value of 3.841.

1. *Were your chi-square statistics lower or higher than 3.841?*
2. *Should you reject HWE for either population?*
3. *If we reject HWE, what does that imply?*

Using the x^2 values we calculated above, compared to the x^2 critical value of 3.841, we can accept that the moths are in HWE, while the Nigerian human population is not. Thus, the Nigerian human population is evolving, at least in reference to this gene. But why is this population evolving? Though we have demonstrated that a population is evolving, we also want to identify the cause of evolution. There are several evolutionary mechanisms that can explain the absence of HWE, indicated by the assumptions that must be true for HWE to prevail.

The HWE assumes that five things are true for the population in question:

1. **There is no natural selection.** Another way to put this is that the fitness of all genotypes is the same.
2. **The population is infinite in size.** While this can never be true for a real population, as we will see, populations can be sufficiently large that the effect of the population size being finite is minimal.
3. **There is no migration.** In particular, individuals from other populations (which might have different allele frequencies from the population under study) are not joining the population in question.
4. **Mating is random.** In other words, an individual with a certain genotype is equally likely to mate with an individual with any other genotype. Note that this is not about sexual selection; sexual selection would fall under the first assumption that there is no natural selection.
5. **There is no mutation.** If an individual's alleles spontaneously change to become different alleles, that can affect allele frequencies. For our purposes, mutation is rare enough that it has little effect on HWE.

4.6 Other evolutionary mechanisms

So far we have focused on the evolutionary mechanism of natural selection, because it is probably the most important mechanism of evolution, it is the most familiar (and often the most misunderstood) mechanism, and it is one of the five major components of Darwin's theory of evolution that we identified in Chapter 2. It is also a very well-studied phenomenon that provides ample opportunities for testing and for illustrating the testability of evolution.

However, natural selection is not the only evolutionary force. Absence of natural selection is only one of five assumptions of the HWE, and therefore there are four other assumptions that could be violated to produce deviations from HWE. In order to avoid leaving you with the mistaken impression that natural selection is the only cause of evolution, let's examine each of these other mechanisms.

4.6.1 Nonrandom mating

HWE assumes that mating is random, meaning that any individual (or any genotype) is equally likely to mate with any other individual (or genotype). **Nonrandom mating** usually is discussed as when individuals preferentially mate with others with the same phenotype or genotype (**assortative mating**) or with individuals with a different phenotype or genotype (**disassortative mating**). One way that nonrandom mating manifests itself is when inbreeding is common. These sorts of nonrandom mating don't actually affect allele frequencies, but they can cause genotype frequencies to deviate from HWE. Nonrandom mating is sometimes confused with sexual selection, which is a form of natural selection, and therefore it pertains to the assumption of HWE that there is no selection.

4.6.2 Mutation

HWE also assumes that there are no **mutations**, or spontaneous changes in a gene. For instance, an individual with the wild-type allele (the typical or ancestral allele in the population) might have a mutation in its sperm or egg that changes the wild-type allele to a mutant version. When that individual has offspring, that obviously can change the frequency of the mutant and wild-type alleles. Mutations, while important for generating genetic diversity over time, are rare events, so their contribution as an evolutionary force is usually rather small as compared with natural selection. Notice that mutations are discussed independent of natural selection. Selective forces don't induce mutations to happen; rather, selection can only act on the existing genetic variation in the population.

4.6.3 Migration

Migration cannot occur for HWE to hold, and migration is an important phenomenon in evolutionary biology. Migration involves more than one population, where individuals from one population can mate with individuals from the other, and vice versa. In many cases, migration actually involves individual organisms physically moving from one population to another, but migration could involve gametes, such as the sperm carried in pollen grains, rather than individuals moving between populations. From the perspective of population genetics, what interests us most is the movement of alleles between populations, or **gene flow**. If the rate of migration between two populations is high, then alleles can readily move from one population to the other, and thus the allele frequencies of the two populations will become more similar. The effects of migration can be seen in the data on rock pocket mice (Figure 4.4); notice that where populations in sandy areas are adjacent to populations on basalt, there tends to be a mix of alleles in the population, even though selection tends to favor the allele related to being cryptic. In Chapter 6, we will be particularly interested in gene flow, not in terms of keeping populations genetically similar, but in terms of how disrupting gene flow can allow two populations to evolve in separate directions, and ultimately lead to them becoming separate species.

4.6.4 Genetic drift

Finally, HWE assumes that population size is infinite, or at least very large. Even a large finite population size will result in something very similar to the predictions of HWE, provided no other assumptions are violated, and a chi-square test would identify the small deviations from the predictions of HWE as not significant. These deviations are due to the effects of chance, similar to how flipping a coin 100 times might not get you exactly 50 heads and 50 tails, although it will probably get you something close to that. In the case of populations, the events are not coin flips, but matings among individuals and the types of offspring they produce. Recall that a Punnett square tells us only the probabilities that the offspring of two parents will have particular genotypes and phenotypes, but, especially among small numbers of offspring, the proportions of genotypes and phenotypes can differ markedly from what the Punnett square predicts. This is also true for the predictions of HWE and the offspring produced in a small population. These effects of chance on the population's allele and genotype frequencies are known as **genetic drift**.

As population size gets smaller, the effects of chance become more pronounced; think of how much easier it is to get a larger deviation from your prediction of 50% heads if you flip a coin only 10 times (or try it a few times to demonstrate this to yourself). Thus, small populations can fluctuate greatly from the predictions of HWE because of genetic drift. This can cause a population with a gene with, say, two alleles to evolve to the point where one allele is lost from the population entirely; the remaining allele is now said to be **fixed** or to have reached **fixation**. Populations that become very small for a period of time are usually characterized by genes with few alleles, few genes with more than one allele, and individuals who are heterozygous for very few genes (also called low **heterozygosity**). Conservation biologists are often concerned about genetic drift, because they deal with species that have small populations, and the loss of genetic diversity often puts those populations at further risk if there are changes in their environments and the population does not have variation that includes individuals able to survive the change.

Table 4.9. Data on genetic diversity of elephant seals reported by Bonnell and Selander (1974) based on proteins. Note the lack of multiple alleles for the proteins of the Northern elephant seal (*Mirounga angustirostris*).

Species	Total number of proteins examined	Number of proteins with more than one form (allele)
M. angustirostris	21	0
M. leonina	18	5

A classic example of genetic drift is the Northern elephant seal (*Mirounga angustirostris*; Figure 4.5), which was hunted heavily in the 1890s, to the point where an estimated 20 or 30 individuals remained. Studies on Northern elephant seal populations, which have increased in numbers in more recent times, have shown that they have very low genetic diversity in comparison to that of the less hunted Southern elephant seal (*Mirounga leonina*) and even compared to genetic data recovered from museum skins of Northern elephant seals collected earlier in the nineteenth century (Table 4.9).

Wildpix 645/Shutterstock.com

Figure 4.5. A male northern elephant seal (*Mirounga angustirostris*).

Note that a small population that later becomes large does not recover alleles lost due to genetic drift. Only mutation or migration from other populations can introduce new alleles. Thus, one effect of a population becoming small is that it goes through what is a called a genetic **bottleneck**. Cheetahs (*Acinonyx jubatus*; Figure 4.6), for

Seyms Brugger/Shutterstock.com

Figure 4.6. A cheetah (*Acinonyx jubatus*).

instance, while not abundant, do not have especially small populations, but they have very low genetic diversity, such that distantly related individuals are nearly genetically identical and could accept skin grafts from each other. This is thought to be due to a bottleneck in the past.

Oleksii G/Shutterstock.com

Figure 4.7. A Tasmanian devil (*Sarcophilus harrisi*).

Another perhaps more bizarre example of the effects of loss of genetic diversity is that of Tasmanian devils (*Sarcophilus harrisi*; Figure 4.7), which have been decimated in parts of their range by a communicable cancer of the face. Cancer is not normally a communicable disease, but the devils are so genetically similar that cancer cells transferred from one individual to another can establish themselves and form tumors without the immune system of the infected devil responding to the cancerous cells. Devil populations were not so small when the disease was discovered to significantly violate HWE, but hunting of devils had reduced their numbers greatly until hunting was banned in the 1940s. This population reduction may have been responsible for the massive loss of genetic variation due to genetic drift.

Note that one consequence of genetic drift is that the direction of change in an allele frequency is random. Thus, two populations of the same species that are small, separate, and otherwise living in similar environments can evolve to become very different through genetic drift. Even though both populations are evolving due to genetic drift, the populations might very likely evolve in different directions. Because change due to genetic drift is random, it also means that alleles can be lost even if they increase fitness, or, conversely, alleles that decrease fitness might become fixed by chance if the effect of genetic drift is strong.

Two populations can also become different even if just one of the two populations is small, such as when a small number of individuals from a mainland population somehow end up on a remote island. In this case, the new population can become genetically different from the source population in two ways. First, the founding individuals of the population may by chance have allele frequencies different from that of the source population. This phenomenon is called the **founder effect**. Second, since the founding population is small, genetic drift might cause it to evolve in directions that the source population does not evolve.

Activity

*The Polynesian cricket (*Teleogryllus oceanicus*) is found on Australia and New Guinea and throughout the Polynesian islands of Oceania. Australia and New Guinea appear to be their ancestral home, and they have spread from there to other islands as far as Hawaii, in some cases presumably by traveling on the boats of humans. Tinghitella et al. (2011) examined the genetic diversity of crickets on Australia and on various islands, including the number of alleles each population had for particular genes.*

1. *Based on what you know about genetic drift and the founder effect, what differences do you predict you would find in terms of the number of alleles for a gene between Australian populations and island populations founded by crickets originally from Australia? Given that more distant islands were likely colonized by crickets from other islands, rather than by Australian crickets, how should more distant islands compare to islands closer to Australia in terms of the number of alleles?*
2. *Figure 4.8 illustrates the results of Tinghitella et al. (2011). How do these compare with the predictions you made above?*

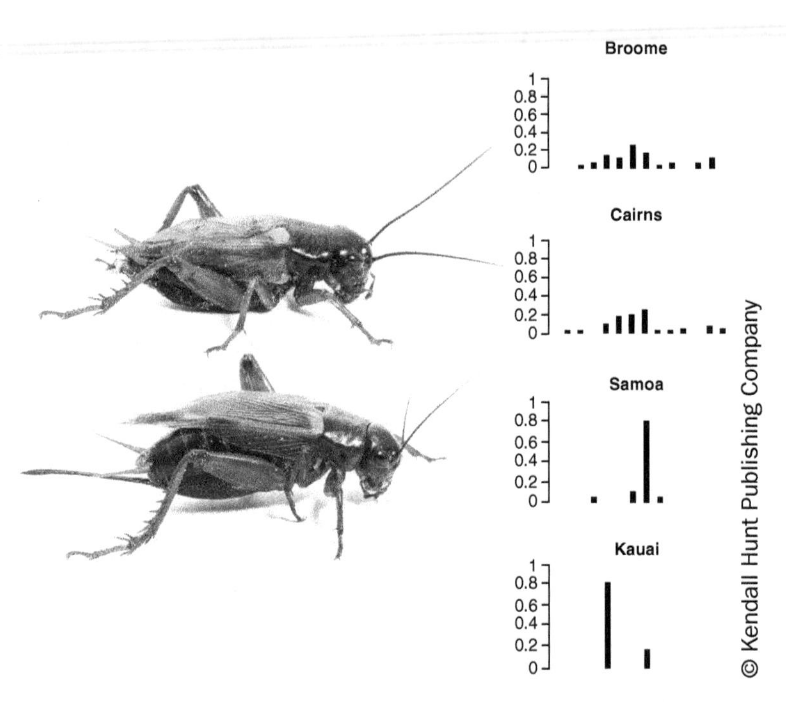

Figure 4.8. Data on allele frequencies for populations of Polynesian field crickets (*Teleogryllus oceanicus*) on Australia and selected Pacific islands, based on Tinghitella et al. (2011). Broome and Cairns are localities in Australia. Samoa and Kauai are Pacific islands, with Kauai farther from Australia than Samoa.

4.7 Selection and humans

We tend to think of natural selection as something that doesn't affect us, but in fact it is very relevant to our lives. When the insecticide DDT was used to control mosquitoes and the spread of malaria, an allele conferring resistance to DDT increased greatly in frequency in mosquito populations; after DDT was banned, the frequency of the allele greatly diminished. A more disconcerting example is the rise in antibiotic-resistant bacteria, such as methicillin-resistant *Staphylococcus aureus* (more commonly known as MRSA), that pose serious health risks to humans. As with DDT and mosquitoes, use of antibiotics can favor selection for resistant strains of bacteria against which we have no effective treatments.

But does selection affect humans directly? Natural selection allows us to explain what was going on in the Nigerian population that we considered in Table 4.7 and the associated activities. The gene we considered in that example was actually the gene for hemoglobin, and the *S* allele is the mutation that leads to sickle cell anemia in individuals homozygous for that allele. We can use the ratios of observed to expected and divide by the highest ratio to calculate the relative fitnesses of the three genotypes; these are given in Table 4.10.

Sickle cell anemia has a high mortality rate, so it is not surprising that the relative fitness of *SS* individuals is so low. We actually might be surprised that the *S* allele is present at all in this population.

Table 4.10. Data from Table 4.7 with blank cells filled, and with calculations of relative fitness. To calculate relative fitness, we can use the ratio of the observed number of individuals (O) to the expected number of individuals (E) for each genotype as a proxy for survival, analogous to the percent recaptured from Kettlewell's experiments in Table 3.2. We then divide these ratios by the highest ratio to get relative fitness.

		SS	*SA*	*AA*	Total	Frequency *S* (*p*)	Frequency *A* (*q*)
Nigerian population	**Observed**	29	2993	9365	12,387	0.123	0.877
	Expected	187.4	2672.4	9527.2			
Ratio O:E		0.155	1.12	0.983			
Relative fitness		0.155/ 1.12 = 0.14	1.12/1.12 = 1.0	0.983/1.12 = 0.88			

The reason the *S* allele is still present at a significant frequency in the Nigerian population lies with heterozygous individuals, who have the highest relative fitness of the three genotypes. Why should individuals with one *S* allele have an advantage over individuals with none? Heterozygous individuals do not get sickle cell anemia (although they are more likely to have children with the condition), but they are also resistant to malaria. Infection of red blood cells of heterozygotes with malaria induces sickling in the infected cell, which destroys that cell but also the pathogen. Thus, heterozygous individuals have an advantage in areas where malaria is present; *AA* individuals are more susceptible to malaria, and *SS* individuals succumb to sickle cell anemia. This is an example of **heterozygote advantage**, and it explains why this population is evolving and not in HWE. In areas where malaria is not present, *AA* individuals and *SA* individuals have the same relative fitness, and selection is stronger against the S allele.

Literature Cited

Bonnell, M L., and R K. Selander. 1974. "Elephant Seals: Genetic Variation and near Extinction." *Science* 184:908–909.

Tinghitella, R M., M. Zuk., M. Beveridge., and L W. Simmons. 2011. "Island Hopping Introduces Polynesian Field Crickets to Novel Environments, Genetic Bottlenecks, and Rapid Evolution." *Journal of Evolutionary Biology* 24:1199–211.

Part III

Macroevolution—Evolution Above the Species Level

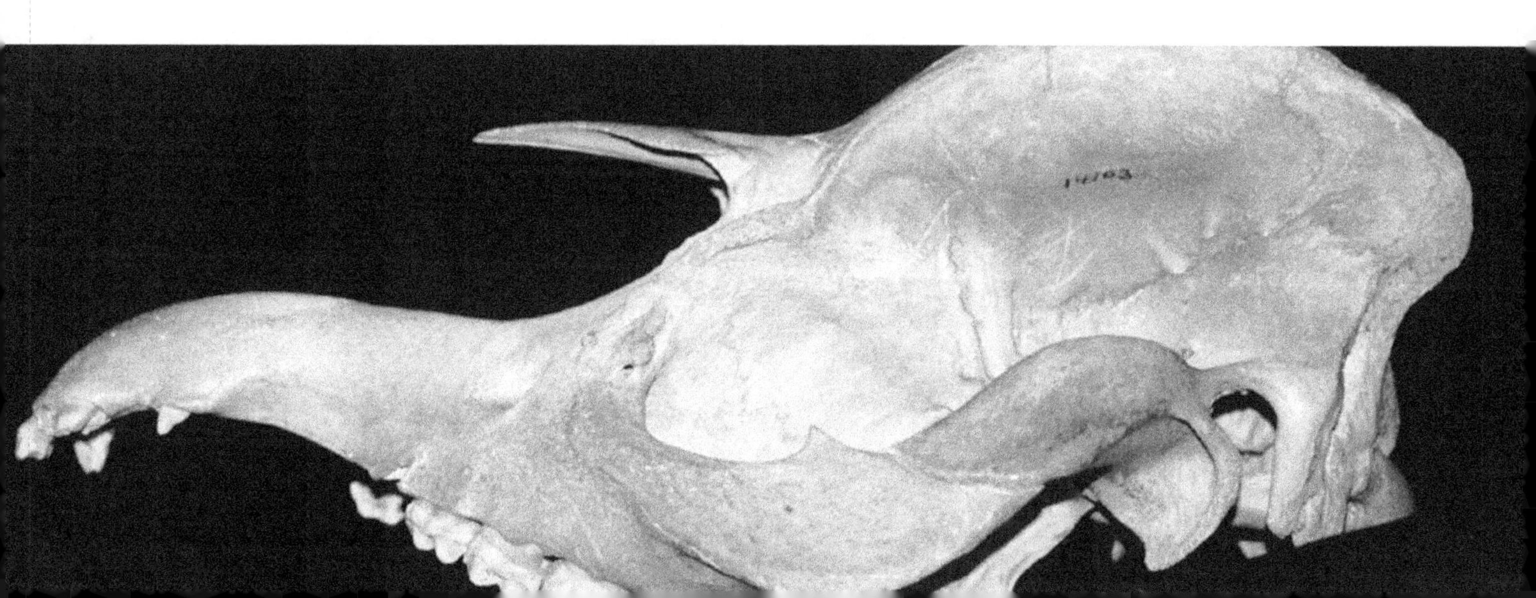

CHAPTER 5 The Predictions of Phylogeny and Macroevolution

In the next two chapters, we are going to look at evolution from the other end of the spectrum of levels of organization. In the previous section, we were primarily concerned with individuals and populations. Evolution at this level is called **microevolution**, and it is the level at which we can truly examine the processes that cause evolution, such as natural selection. We now want to to turn to the consequences of evolution, the phenomena that we identified in our five theories model as multiplication of species and common descent. This means moving away from microevolution and focusing on macroevolution.

Macroevolution, or evolution above the level of species, might seem like an unlikely area for testable hypotheses. Natural selection and changes in genes over generations might not happen quickly, but they happen fast enough to be measured, in many cases in less than the lifetime of a human. Patterns of evolution across species, on the other hand, are generally measured in geologic time. How can scientists test phenomena that happened millions of years in the past?

Although we can't make direct observations of macroevolution occurring in the past, we can still make predictions of what its effects should look like today. For instance, we can make predictions about the geographic distribution of species based on the notion of speciation, which we will discuss in Chapter 6, or about the number of differences between the genes of different species based on how closely or distantly they are related. The key to testing macroevolution is matching patterns like geographic distribution to patterns of evolutionary history.

In this chapter, we'll investigate some of the tools used by evolutionary biologists to study phylogeny and macroevolution, and we'll apply these tools to some macroevolutionary questions to demonstrate how they make testable claims.

5.1 Phylogeny

Everything that we observe about organisms is potentially explained by their evolutionary history, or **phylogeny**. For example, if we are trying to explain why two species are similar in some way, an important explanation to consider is that they inherited that similarity from a common ancestor. In fact, it is important to account for phylogeny whenever we are trying to understand macroevolution. For instance, if I am interested in something that I believe is related to small body size in mammals, but every small mammal that I study is a rodent, I have to account for the possibility that the phenomenon I am studying is not so much related to being small as it is to being a descendant of the common ancestor of rodents.

5.1.1 Reading trees

When we investigate phylogeny, we are interested in **phylogenetic relationships**, or how different groups of organisms are related to one another. We typically represent the pattern of relationships in a tree-like diagram. There are several terms used for these tree-like diagrams, including phylogeny, **evolutionary** or **phylogenetic tree**, and **cladogram**. It is important to understand what information is contained in a cladogram, and how to read it.

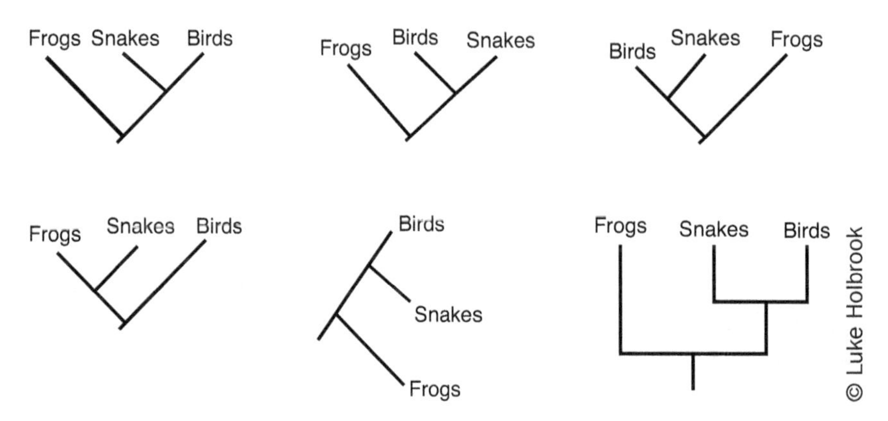

Figure 5.1. Examples of phylogenetic trees drawn in different ways. Only one tree shows a different set of relationships from the others. Can you tell which one?

Figure 5.1 shows some typical ways that phylogenetic relationships among organisms are represented in books and scientific articles. All but one of the trees depicted in this figure are showing the exact same information, namely which taxa are more closely related to each other than to the others. The term **taxon** (pl. **taxa**) refers to any species or group of species that we might represent on a tree. Each of the groups represented at the tips of the branches is a taxon; in fact, these are often called **terminal taxa**, because they are at the ends (or terminals) of the branches. Note that we could identify a group of terminal taxa that could also be a taxon. For instance, in Figure 5.3, placental mammals, marsupials, and monotremes are all mammals, so they together would constitute the taxon Mammalia. Taxa are ultimately what go into our classifications, which we will discuss later in this chapter.

Note that, besides the terminal taxa, each tree consists of lines that we call **branches** and points where the branches connect, called **nodes** (Figure 5.2). Nodes represent a couple of things. Most importantly, they represent **hypothetical ancestors**; in other words, taxa that share a common ancestor also share a node. We don't identify that ancestor as, say, a specific fossil, but we don't need to. It's similar to how, if I introduced you to my brother or sister, you would be able to infer the existence of our mother and father that we share, even if you have never met our parents, and that I share these ancestors with my siblings but not with anyone else. The taxa that share a specific node share a common ancestor that is unique to them, at least relative to the other taxa on the tree.

Notice that a node typically gives rise to two branches, and some nodes arise from other nodes. A node therefore also represents a point when that ancestor gave rise to two descendant

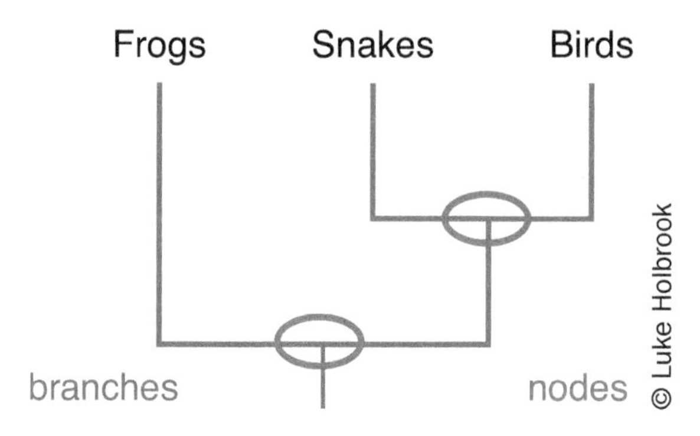

Figure 5.2. A cladogram with nodes (circled in red) and branches (the blue lines).

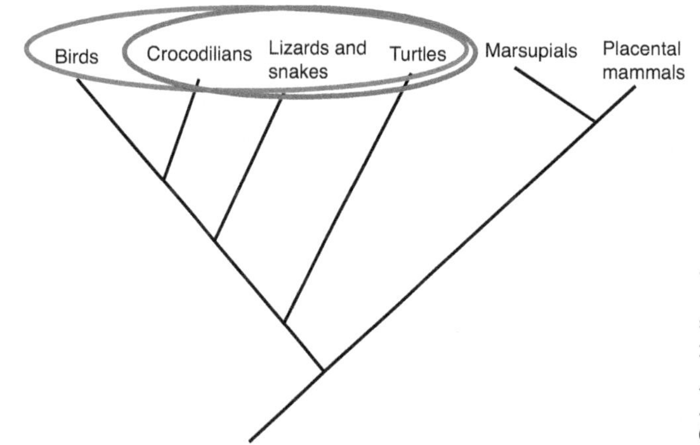

Figure 5.3. A phylogeny of groups of amniotes, a group of vertebrates including mammals, birds, and "reptiles." The red oval is an example of a monophyletic group. The blue oval represents a group that is not monophyletic.

lineages; this splitting event essentially represents an instance of **speciation**, the process by which new species arise, which we will discuss in Chapter 6.

Nodes also tell us something about the timing of evolutionary events. We can think of the ordering of nodes as reflecting which ancestors are relatively older or younger. Going back to the family analogy, if I introduced you to my brother and my cousin, you would be able to infer that my brother and I share a more recent ancestor (my parents) with each other than the ancestor we share with our cousin (our grandparents). Note that my siblings and I share many ancestors, including our parents and our grandparents, but our parents are the only ancestors that we uniquely share, and they are the most recent ancestors that we share.

Similarly, the taxa on a cladogram often share more than one ancestor, indicated by the nodes on the tree. For instance, in Figure 5.3, marsupials and placental mammals share an ancestor that is unique to them and not shared by the other groups on the tree. Marsupials and placentals also share an older ancestor, which is also shared by monotremes, and an even older ancestor that they share with all of the taxa on the tree. If we want to specify that we are talking about the ancestor that is uniquely shared by placental mammals and marsupials, we can refer to that as the **most recent common ancestor** of placental mammals and marsupials.

As mentioned before, the basic information that you get from a cladogram is which taxa are more closely related to each other than to the others in the tree. So, you can see in Figure 5.3 that placentals and marsupials are more closely related to each other than to anything else, and you can also see that monotremes are more closely related to the placental-marsupial group than to the remaining taxa on the tree. Another way to say this is that placentals and marsupials share a unique common ancestor, as do placentals, marsupials, and monotremes. Contrast this with the common ancestor that turtles, lizards and snakes, and crocodilians share; we can't say that they share a unique common ancestor, because birds also share this ancestor.

In biology, a taxon is considered to be all of the descendants of a particular ancestor, indicated by a node, not simply a selection of that node's descendants. A group that includes all of the descendants of a particular node is called a **monophyletic** group. Thus, based on our tree, Mammalia (placentals, marsupials, and monotremes) is an example of a monophyletic group, whereas "reptiles" (turtles, lizards and snakes, and crocodilians) would not be, because it ignores the fact that the most recent common ancestor that these groups share is also shared by birds.

Activity

Identify all of the monophyletic groups represented in the tree in Figure 5.3.

Because cladograms tell us who is more closely related to whom, we can make statements about relationships among subsets of taxa on a larger tree. If we select a set of taxa from the tree in Figure 5.3, we can draw a new tree with only those taxa, based on the tree in Figure 5.3. For instance, if I was only considering birds, mammals, and turtles, based on the tree in Figure 5.3, I could draw a new tree showing that turtles and birds are more closely related to each other than they are to mammals (see Figure 5.4). Similarly, if someone drew a tree relating birds, mammals, and turtles in some way, I could determine whether that tree was consistent with the relationships depicted in Figure 5.3.

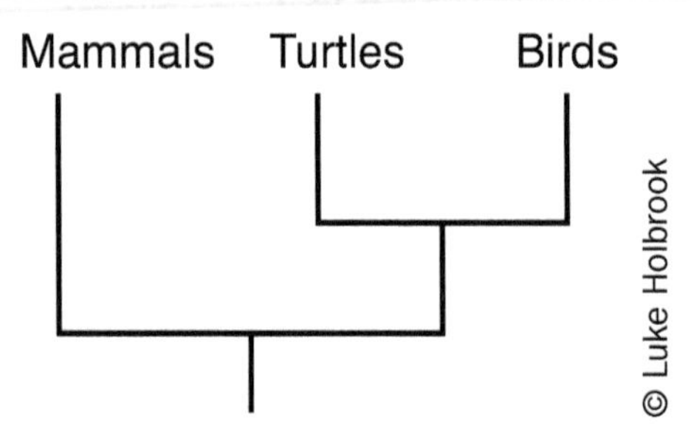

Figure 5.4. A tree of turtles, birds, and mammals based on the tree in Figure 5.3.

Activity

Choose three of the terminal taxa from Figure 5.3, perhaps at random, and draw a tree of just those three taxa that is consistent with the relationships in Figure 5.3. Can you give an example of a tree for these taxa that would be inconsistent with the relationships in Figure 5.3?

5.2 Phylogenetic analysis

One of the most important tools for investigating macroevolutionary patterns is **phylogenetic analysis**, the study of evolutionary relationships among species. Phylogenetic analysis allows us to evaluate different hypotheses about how species are related. A cladogram is a hypothesis of relationships: it represents one possible way that all of the taxa in the tree are related to one another. In fact, for any given number of taxa, there is a finite number of possible cladograms, and therefore a finite number of hypotheses, one of which represents the true relationships among the taxa. To say that it is finite is not to say that the number is necessarily small; there are three possible cladograms for three taxa, 15 for 4 taxa, 105 for 5 taxa, and more than 34 million for 10 taxa. Each additional taxon greatly increases the number of possible trees: the number of possible trees for 50 taxa is greater than the number of atoms in the universe. The question then is: How can we choose the best hypothesis from among the many possibilities?

Activity

Draw all of the possible trees for the following three taxa: frogs, snakes, and birds.

5.2.1 Characters

The main way we test phylogenetic hypotheses is to essentially use them to predict how traits should be distributed. Closely related taxa should inherit similar things from their common ancestor; thus, we expect taxa with similar traits to be closely related. Conversely, we wouldn't expect distantly related taxa to have some unique similarity that they do not share with other closer relatives, at least not based on common ancestry. Of course, we don't expect every trait to be the same for taxa sharing a common ancestor; it is possible that an ancestral trait will change or be lost in a descendant taxon, or that the same trait might evolve independently in two or more different taxa. As we'll see, similarities among distantly related taxa can occur due to independent evolution, but we expect to find similarities among closely related taxa more often than among distantly related taxa.

Essentially, what we do in phylogenetic analysis is record the **states** of various **characters** for the taxa in our analysis, and, based on the distribution of character states among taxa, determine which phylogenetic hypothesis fits that distribution best. Characters can be just about any aspect of an organism, like the traits we discussed in Chapters 3 and 4, but important criteria for good characters are that they are heritable, that they usually do not vary greatly within a terminal taxon, and that they show at least some variation among the taxa in our analysis. We can record the data on character state variation in a character-taxon matrix; an example of such a matrix is given in Figure 5.9. Essentially, this is a table that gives a score for each taxon based on the observed state of each character.

Historically, the characters used in phylogenetic analysis at first were almost exclusively anatomical traits, and for analyses that include fossils these are still important. In recent decades, DNA sequences have allowed us to directly assess genetic similarities across species from which we can obtain sequences, typically living species, but in some cases fossils.

A Closer Look: DNA as Phylogenetic Information

Most students are familiar with **DNA**, but it's useful to review some aspects of it here, since the structure of this molecule is critical to understanding just about all of biology, and in particular it is important for our discussion of phylogeny.

DNA structure

DNA stands for **deoxyribonucleic acid**. This molecule is the molecular basis of heredity. It is part of an important class of biological molecules called **nucleic acids**. It is also a polymer, meaning that it is made up of a series of subunits, like the links in a chain; these subunits are called **nucleotides**. DNA actually consists of two such chains linked to each other in parallel; in fact, the chains are **antiparallel**, because they actually run in opposite directions in terms of the direction in which the ends of the nucleotides are "pointed." The shape is analogous to a ladder, except that it is also twisted into a spiral shape, forming the iconic **double helix** shown in Figure 5.5.

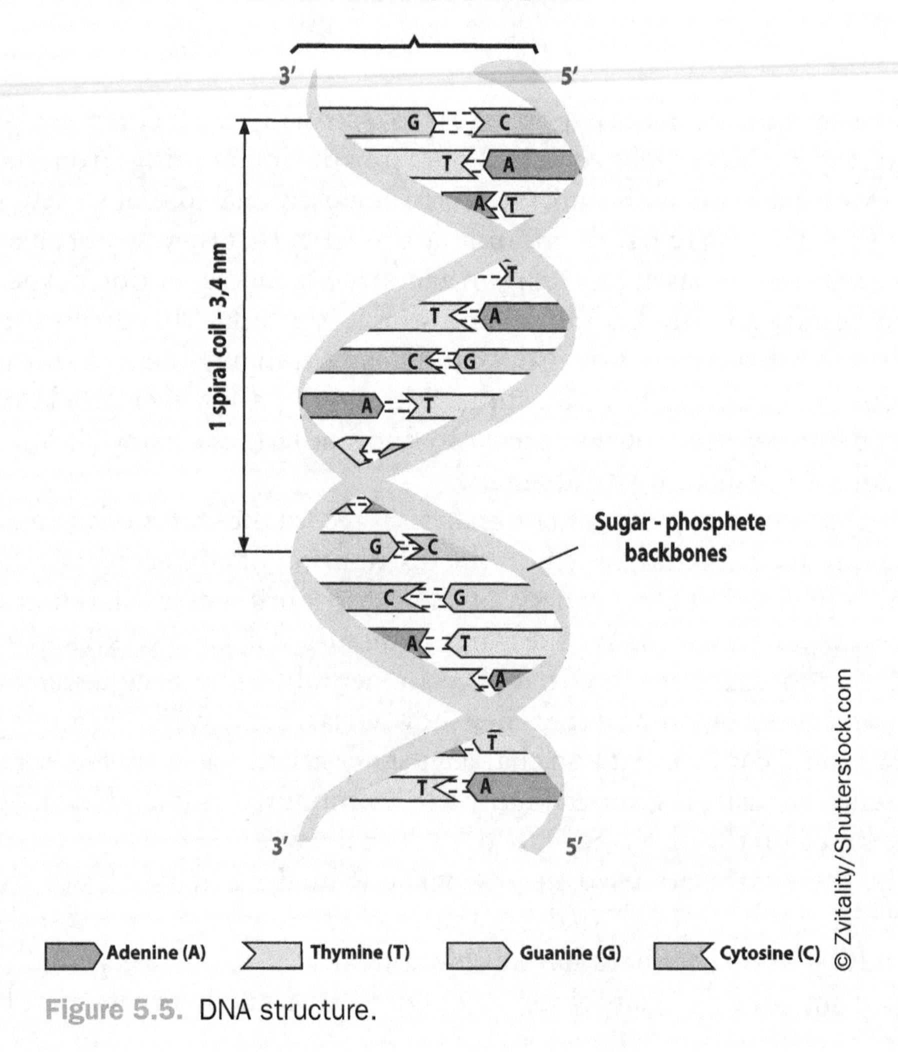

Figure 5.5. DNA structure.

The structure of nucelotides determines the important properties of DNA. Each nucelotide consists of three components: a **sugar**, a **phosphate group**, and a **nitrogenous base** (Figure 5.5). The sugars and phosphates are the same in every DNA nucleotide, and they essentially form the links in the "chain" that forms each of the two strands. Each sugar-phosphate portion has two different ends, referred to as five prime (5′) and three prime (3′), and nucleotides in each strand link together by bonds between the 5′ end of one nucleotide and the 3′ end of the next. Because the two strands are antiparallel, they run in opposite directions in terms of which end is 5′ and which is 3′.

The nitrogenous bases can be different from nucleotide to nucleotide, and they can connect to nitrogenous bases on the opposite strand by means of **hydrogen bonds**, which are easier to break than the covalent bonds that hold together the rest of the molecule. The weakness of hydrogen bonds is important in DNA function, as we will see.

There are four types of nitrogenous bases in DNA: **adenine**, **cytosine**, **guanine**, and **thymine**. Thus, there are four types of nucleotides in DNA, which we typically identify using the letters ***A***, ***C***, ***G***, and ***T***. So, a specific single strand of DNA will consist of a specific sequence of nucleotides that we can represent with letters. For instance, we could have part of a strand composed of the following sequence of nucleotides: *ATTAGCGCCCTAAG*. For the sake of the example, let's assume that the 5′ end is on the left and the 3′ end is on the right.

There is one more important property of the nitrogenous bases: each base binds only with one of the other three, as illustrated in Figure 5.5. So, *A* binds with *T*, and *C* binds with *G*; another way of saying this is that *A* and *T* are **complementary**, as are *C* and *G*, that means if I know the base sequence of a single strand of DNA, I can determine the other (complementary) strand. So, the strand *ATTAGCGCCCTAAG* from the last paragraph would have the following complementary strand: *TAATCGCGGGATTC*. Of course, this strand is running antiparallel to the other, so if we wanted both strands to read as having the 5′ end on the left, we would need to flip the complementary strand around: *CTTAGGGCGCTAAT*.

DNA replication

The structure of DNA allows it to perform two critical functions. First, it can be copied, through a process called **replication**; new copies produced by replication facilitate the passage of DNA to new cells, or to new offspring. Second, the nucleotide sequence can be "read" to determine the composition of a protein being constructed by a cell. **Proteins** are another important class of biological molecules that perform many of the molecular functions that determine phenotype, either as parts of cell structure or as enzymes that catalyze specific chemical reactions. Thus, the genetic information in DNA can be used to determine the organism's phenotype.

DNA replication, illustrated in Figure 5.6, involves first breaking the hydrogen bonds between the two strands, analogous to "unzipping" the double helix into two single strands. Next, the two separated strands serve as templates for the construction of the complementary strands of what are now two new DNA molecules. In essence, the original strands bind with new nucleotides to form a new second strand on each; because nitrogenous bases only bind with a complementary base, the new strand should have a composition exactly like that of the complementary strand from the original molecule. Thus, when DNA replicates, it results in a molecule that has one strand from the original molecule and one strand that was constructed anew; DNA replication is thus termed **semiconservative**, because it conserves half of the old molecule in each copy.

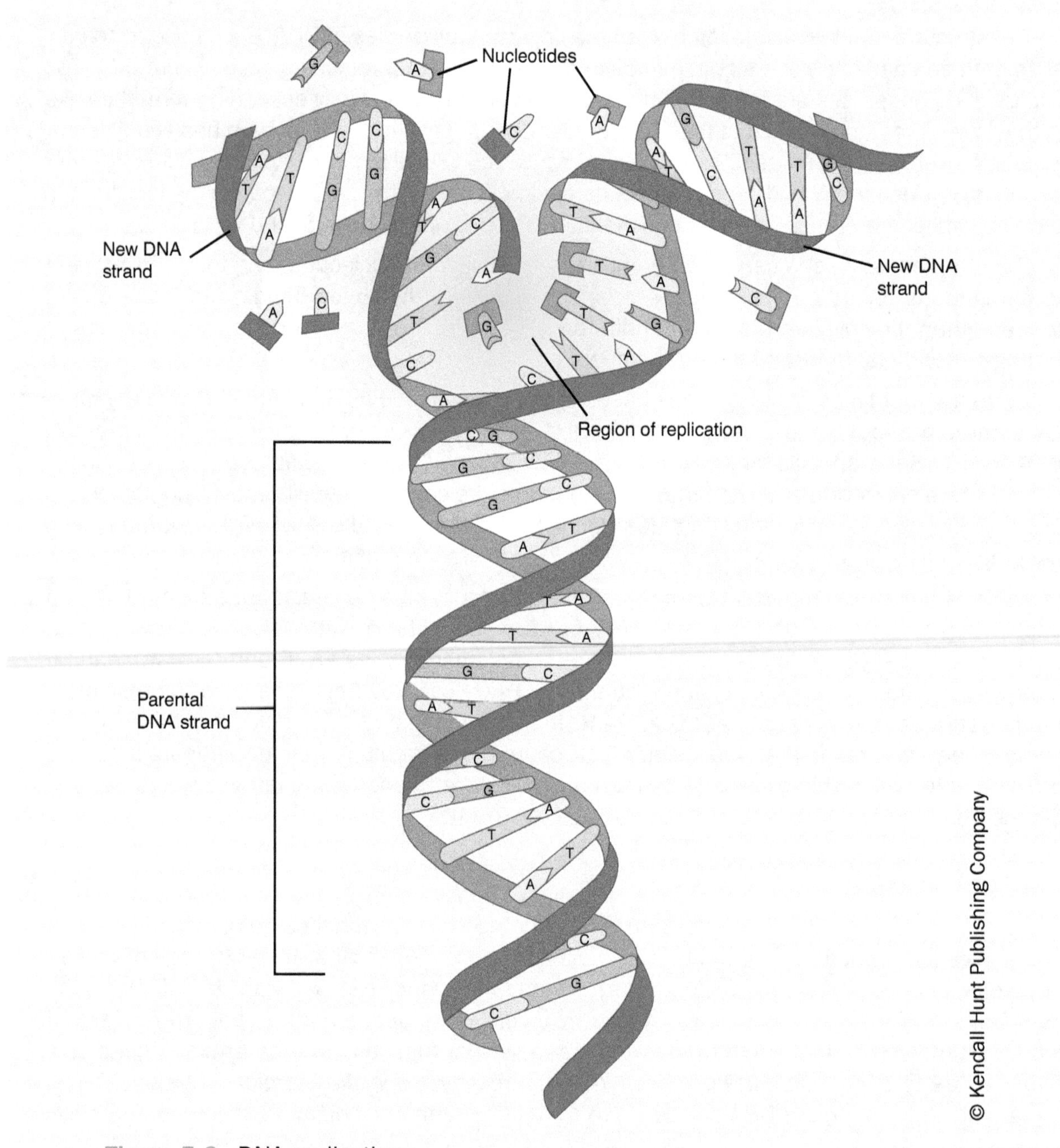

Figure 5.6. DNA replication.

Once DNA is replicated, copies of an organism's genes can be transferred to new cells, such as those generated as the organism grows. They can also be packaged in gametes (sperms or eggs) to produce offspring that will inherit those copied genes.

DNA transcription and translation

Genes are ultimately specific sequences of DNA, and alleles are simply gene sequences that differ in some way, possibly in a way that causes a difference in phenotype. But how does a DNA sequence determine a phenotype? First, we need to understand that phenotypes ultimately translate to proteins. Proteins are important for structures in our bodies, but, perhaps more importantly, they provide the enzymes that catalyze the various reactions that ultimately determine our phenotype. This includes the enzymes that control which genes are "turned on" or "turned off," such that proteins can facilitate very complex processes in an organism, and thus can determine a wide range of complex phenotypes. A protein has a number of features, but the most important one for our purposes is that, like DNA, it is a polymer, made up of a sequence of units called **amino acids**. There are 20 kinds of amino acids that contribute to the makeup of the proteins involved in the processes of living things.

The process of going from a sequence of nucleotides to a sequence of amino acids involves two general steps. The first is **transcription**, illustrated in Figure 5.7, essentially copying the sequence of the DNA to another molecule, **RNA** (or, more specifically, **messenger RNA** or **mRNA**). RNA is very much like DNA, except that it is typically single stranded, its nucleotides have a different kind of sugar (ribose instead of deoxyribose), and its four types of nitrogenous bases include one called **uracil** in place of thymine. Like thymine, uracil is complemetrary to adenine.

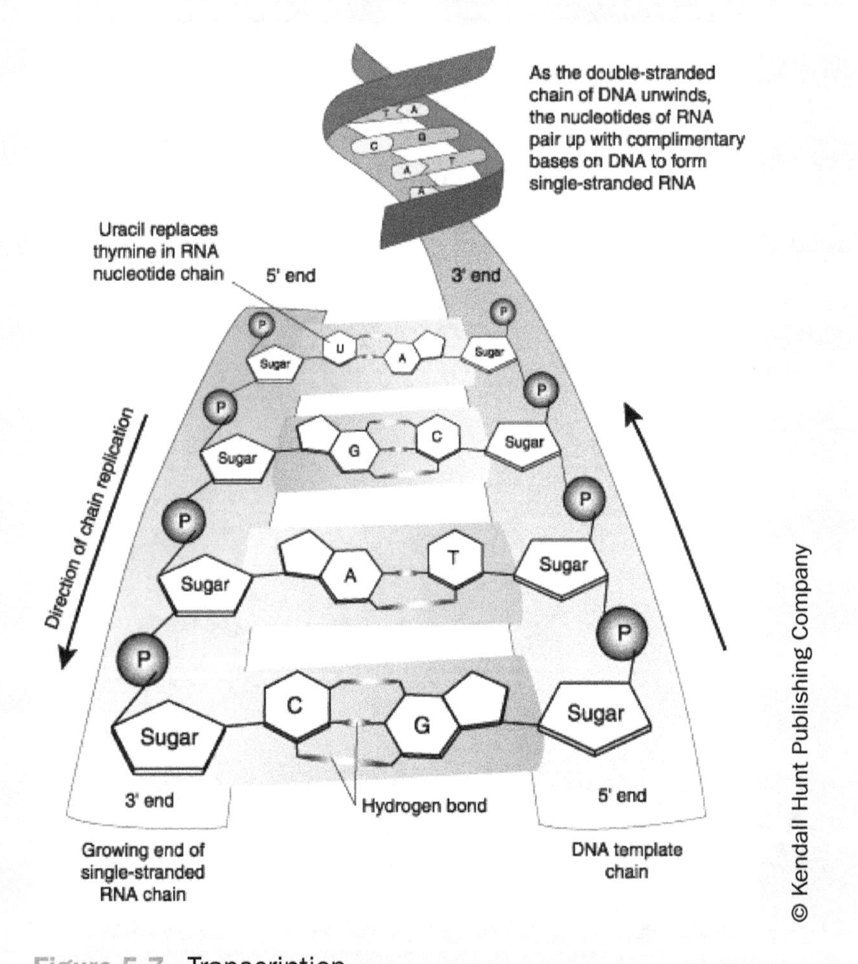

Figure 5.7. Transcription.

In transcription, as in replication, the DNA is "unzipped," but only at the site to be transcribed. Nucleotides of RNA bind to the complementary bases on the DNA strand. Thus, mRNA is complementary to the gene being transcribed. The mRNA then travels to the site of protein synthesis, an organelle called the **ribosome**, which is also composed of RNA (called **ribosomal RNA** or **rRNA**).

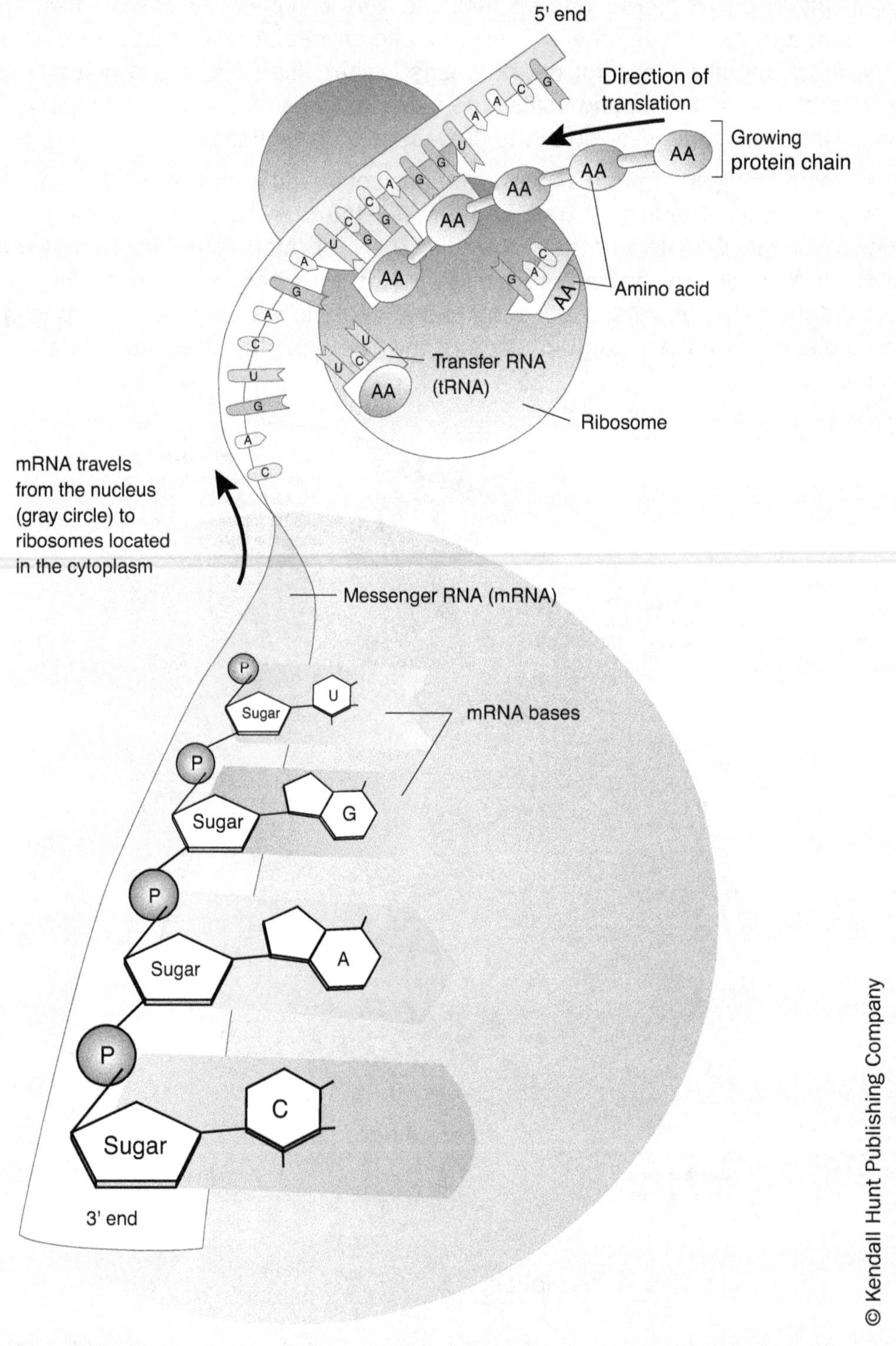

Figure 5.8. Translation.

At the ribosome, the mRNA undergoes the other major step, **translation**, illustrated in Figure 5.8. Essentially, the nucelotide sequence of the mRNA is matched to a corresponding sequence of amino acids, forming the primary structure of the protein. Since there are only four types of nucleotides and there are 20 kinds of amino acids, the specific amino acid to be added to the protein sequence is determined by a sequence of three nucleotides, called a **codon**. The four types of nucleotide allow for 64 different three-nucleotide combinations, so there is some redundancy, where multiple codons code for the same amino acid.

During translation, the mRNA is "read" at the ribosome as a series of three-nucleotide codons, and the protein is constructed as a series of amino acids corresponding to the sequence of codons. Thus, what was originally encoded as a sequence of DNA nucleotides has been used to determine the sequence of amino acids to form a protein that contributes to the phenotype.

DNA and phylogeny

What we've discussed so far explains how DNA forms the molecular basis for genes, how it can be copied (and therefore inherited), and how it can determine phenotype. So, how does this translate into information that we can use in phylogenetic analysis? DNA sequences do vary among organisms and among taxa; for instance, at a particular point in a gene sequence where organisms in one taxon have a *G*, those of another taxon might have a *T*. We can also look at those differences and, because we know how codons translate to amino acids, determine if this will result in a change in phenotype.

Perhaps, more striking than the variation in sequences that we see among organisms is the conservation of gene sequences that we also observe. One reason we can compare gene sequences among organisms is that they are often very similar across a wide range of taxa, and closely related taxa, as we might expect, have very similar gene sequences. One expression of this is the commonly cited similarity of humans and chimpanzees, where upwards of 98% of our DNA sequences are the same.

This combination of variation and conservation is very accommodating for phylogenetic analysis. Conservation of genes allows us to identify the same gene in different taxa, determine their sequences, and line up their sequences, so that we can compare the equivalent positions in the sequence for each taxon; this process of lining up similar sequences is called **alignment**. The variation in some nucleotide positions across taxa can be used as characters for our phylogenetic analysis.

Mutation

Why, then, does this variation exist, and why isn't there more of it? The description of DNA structure and function above is greatly simplified, and one thing that was left out is the imperfection of the process. Replication is not always perfect, and occasionally mistakes are made, resulting in **mutations**. Mutations can be small, involving only a single nucleotide (a *G* added instead of a *T*, an extra nucleotide inserted, or one omitted), these are called **point mutations**. Mutations can also affect a larger sequence of multiple nucleotides (deletion or insertion of a whole sequence of nucleotides, or flipping the order of a sequence around). In either case, the effect of a mutation on phenotype can vary in significance; even changing a single nucleotide can change an important amino acid in a protein, or shift the reading of the resulting mRNA "downstream" of the mistake, a type of change called a **frameshift**.

If a mutation results in a significant phenotypic change, it may disappear due to selection; most mutations are likely to alter phenotypes in a way that reduces fitness. Only rarely will a mutation actually increase fitness, at least in some circumstances, and thus spread in the population due to selection.

It is also possible that a mutation in a DNA sequence will result in no phenotypic change. It could be that the mutation results in a codon that codes for the same amino acid as it did before the mutation; these are called **silent mutations**. There are also many parts of a DNA molecule that have little to no effect on phenotype. They may be parts whose RNA complements are snipped out of the mRNA during the "processing" that occurs before reaching the ribosome. Some genes may no longer function at all (often called **pseudogenes**), and therefore changes in their sequences are never transcribed. There are also large amounts of **junk DNA** that may have no function at all (or these sequences may be pseudogenes, or perhaps we simply don't yet understand their function).

5.2.2 Determining the cladogram that best fits the data

There are several types of criteria that we can use to determine which of our many possible phylogenetic hypotheses best fits the data. For our purposes, we will use a criterion called **parsimony**. Parsimony is essentially the first criterion that was employed for finding the best supported tree, and it is still used in a number of contexts, particularly with morphological data. However, most studies that analyze molecular data use **model-based** methods, specifically maximum likelihood or Bayesian inference, which use criteria for determining the best supported tree that differ from that of parsimony, and which have certain advantages over parsimony, particularly with molecular data. We will introduce some of these model-based methods later, but here we will illustrate phylogenetic analysis using parsimony, as it is a fairly straightforward approach that is easy for students to grasp.

Parsimony as a general concept is a version of "Occam's Razor": the simplest explanation is the best one. In this case, the "simplest" explanation means the hypothesis that requires the fewest evolutionary events or transformations. A transformation would be where at some point on the tree, based on the observed distribution of character states, we would infer a change from one state to another. Note that, for a given tree, a change from one particular state to another need not happen only once, and in some cases for a particular tree we must infer that a particular transformation happened more than once.

In order to assess the number of changes, one thing we need to know is what the states are in the common ancestor of all the taxa, and thus what the initial direction of change will be. For example, if some taxa have long legs and others have short legs, did the ancestor of all of these taxa have short legs that became longer in some of the taxa, or the other way around? The most common way that we deal with this in phylogenetic analysis is to use an **outgroup**. An outgroup is a taxon (or a set of taxa) that lies outside of the group that we are studying (termed the **ingroup**). For instance, for our example of frogs, snakes, and birds, we could use a fish as an outgroup. Note that the outgroup is not the ancestor of the ingroup taxa;

rather, the only feature an outgroup must have is that the ingroup taxa share a more recent common ancestor with each other than they do with the outgroup.

Once we select an outgroup, any character states that the outgroup shares with any members of the ingroup are considered to be **ancestral**; these states preceded the other character states, in an evolutionary sense, and they were present in the ancestor for the ingroup. In other words, we would expect that the ancestor shared by the ingroup and the outgroup would have character states shared by the outgroup and by at least some of the ingroup. All other character states are considered to be **derived**; in other words, they evolved as a change from the ancestral condition. For example, if the number of toes varies in our ingroup from one to five, and our outgroup has five toes, five toes would be considered ancestral. When we make determinations about ancestral and derived states, we are determining what we call the **polarity** of those states. In actual practice, the outgroup is used to root the tree, and by doing so we determine the direction of character state changes.

Activity

Examine the character-taxon matrix in Figure 5.9. using the trout as the outgroup, determine the ancestral and derived conditions for each character.

TAXON	FEATHERS	LIMBS/FINS	SHELLED EGG	WATER-PROOF SKIN
Trout	N	Y	N	N
Frog	N	Y	N	N
Bird	Y	Y	Y	Y
Snake	N	N	Y	Y

Figure 5.9. Character taxon matrix with outgroup. In this example, the outgroup is the trout. "N" and "Y" represent whether a taxon has ("Y") or lacks ("N") a particular feature.

We now have looked at and recorded the variation in some set of characters for a set of taxa. We've selected an outgroup, which allows us to determine polarity. How can we now use this information to evaluate the different possible hypotheses? To do this, we need to be able to map the characters onto the tree in a manner that allows us to count how many transformations the tree implies.

Figures 5.10 and 5.11 illustrate an example of mapping a character onto different trees. Ideally, each character transformation occurs only once, because all of the taxa that have a given derived state are more closely related to each other than to any other taxa and inherited this state from their most recent common ancestor. Note that, in this case, a shared derived character state is characteristic of a monophyletic group. In fact, what we are effectively doing in phylogenetic analysis is looking for evidence to support specific monophyletic groups, and the evidence comes from shared derived characters.

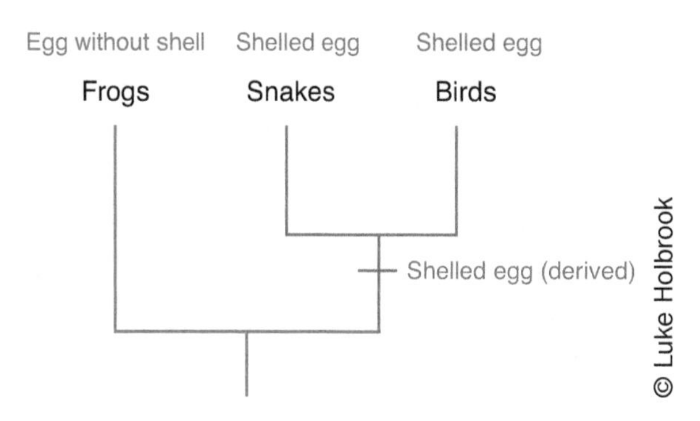

Figure 5.10. Mapping characters onto trees.

Of course, this ideal is not what we always observe. Many character states exhibit **homoplasy**, meaning that they have evolved multiple times. The term homoplasy is used in contrast to the term **homology**, which is similarity due to common ancestry. Homoplasy can be exhibited in two ways on a tree. The simplest is where the same derived state evolved multiple times in different lineages. In this case, the derived state is not **homologous** in the taxa that possess it; in other words, their similarity in this regard is not due to common ancestry, and the derived state evolved independently in each lineage. The second way is where the ancestral state evolves again from a derived state; we call this a **reversal**. Note that both kinds of homoplasy require more transformations—or "**steps**" on the tree—than the ideal condition of all the members of a monophyletic group inheriting the derived condition from a common ancestor. Note that in some cases it may not be possible to determine whether a derived state evolved independently or a reversal occurred, because both types of transformations would have the same number of steps on a given tree. As we will see, this does not hamper our ability to assess how well the data fit the tree.

Activity

Map the transformations for each of the characters in the character-taxon matrix in Figure 5.9 onto the three possible trees for frogs, snakes, and birds.

1. *How many transformations does each tree have?*
2. *Which tree is the "shortest," or has the fewest steps? Which is the longest?*
3. *Parsimony is the criterion that the simplest explanation is the best one. Based on this, which tree is the simplest? Which is the best supported hypothesis?*

The shortest tree turns out to be the one uniting birds and snakes in a monophyletic group. Note that the mapping of gaining feathers and losing legs essentially had no effect on the results; these derived states contributed one step each on all trees, because they are

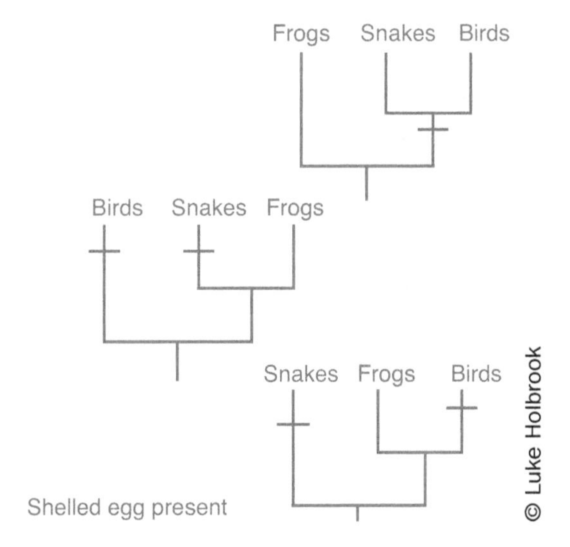

Figure 5.11. Shelled egg character mapped onto the three possible trees for these three taxa.

found only in birds and snakes, respectively. The character states that made a difference are presence of a hard-shelled egg and presence of waterproof skin, because these were shared derived characters of birds and snakes. In the shortest tree, these states evolved only once, whereas in the others they had to evolve twice (or evolve once and reverse back to the ancestral state). Note that whether we interpreted these characters as evolving independently or as reversals contributed the same number of steps to the length of the tree.

A Closer Look: Model-Based Methods and Criteria Other Than Parsimony

Parsimony is a relatively easy criterion to understand and to apply, but it has some limitations. It does not lend itself particularly well to statistical tests, so it is difficult to determine whether the shortest tree is significantly better than the next shortest tree. It also is prone to an effect called **long branch attraction**, particularly when we consider the phylogenetic analysis of sequence data. At every position in a sequence of DNA, there are only a few possible mutations: for instance, if a position ancestrally has a nucleotide with cytosine (*C*), the possible mutations are (1) a change to *A*, *G*, or *T*; or (2) deletion of that position (represented in matrices as "-" for "gap"). If that position mutates more than once, it is quite possible that it could change back to the ancestral state. This possibility is more likely with the passing of more evolutionary time. Thus, two distantly related taxa might have the same nucleotides at the same position not because they inherited them from a common ancestor, but because that similarity evolved independently over the long period of time since the two taxa diverged from their common ancestor. The tendency for parsimony to treat these taxa as closely related based on this similartity is long branch attraction.

Maximum likelihood or **ML** was developed as an alternative to parsimony that is not prone to long branch attraction. This method essentially looks for the tree for which the data would have the highest likelihood. To put it another way, ML asks for each tree, "What is the probability that we would end up with the sequences that we have for these taxa, given this specific tree?" The tree that gives us the highest likelihood for the data is considered to be the best supported.

ML requires a model of evolution for nucleotides (Figure 5.12). The simplest model (known as the Jukes–Cantor model) basically says that the probability of mutating a nucleotide into any other nucelotide is the same for all nucleotides. Of course, this is not necessarily the biological reality. Another model is called the **two-parameter model**, because, instead of one probability for all changes as in Jukes–Cantor, this model separates possible changes into two kinds, each with their own probability, or parameter. The rationale for the two-parameter model is that the four nucleotides fall into two classes based on the number of rings in their nitrogenous bases. Adenine and guanine are **purines** and have two rings, whereas cytosine and thymine (and uracil in RNA) are **pyrimidines** with a single ring. Mutations that substitute a purine for a purine or a pyrimidine for a pyrimidine are called **transitions.** Mutations that substitute a purine for a pyrimidine or vice versa are called **transversions**. Transitions are chemically more likely than transversions, because transitions exchange more structurally similar molecules. The two-parameter model calculates separate probabilities for transitions and transversions. The model that is probably used the most is the most complex and is called the **general time reversible model**, or **GTR**. GTR essentially says that each pair of nucelotides has its own probability of changing from one to the other, although the probability is the same regardless of the direction of the change. Thus, for GTR, the probability of changing from *C* to *G* is the same as the probability of changing from *G* to *C*, but it is different from the probability of changing between *C* and *A* or *C* and *T*. This results in six different parameters that correspond to the probabilities of the different possible mutations.

A

	A	C	G	T
A	-	α	α	α
C	-	-	α	α
G	-	-	-	α
T	-	-	-	-

B

	A	C	G	T
A	-	α	β	α
C	-	-	α	β
G	-	-	-	α
T	-	-	-	-

C

	A	C	G	T
A	-	α	β	γ
C	-	-	δ	ε
G	-	-	-	ζ
T	-	-	-	-

Figure 5.12. Models of sequence evolution for maximum likelihood estimation of phylogeny. Each box in the table indicates the parameter for changes from one nucleotide to another (with the assumption that it would be the same probability in either direction). (A) The Jukes–Cantor model, with one parameter for all changes; (B) the two-parameter model, with different parameters for transitions (β) and transversions (α); (C) the general time reversible model, with each pair of different nucleotides having its own parameter.

ML is amenable to certain kinds of statistical tests, but typically the results are reported as the single tree with the highest likelihood. More recently, another kind of analysis that uses ML has been adopted more widely, because it allows for assessing confidence in a set of trees. **Bayesian inference** (**BI**) is based on Bayes' theorem, which can be expressed as follows, where *A* is a hypothesis, *B* is new data or evidence, *P* is "the probability of," and the vertical bar (|) reads as "given":

$$P(A \mid B) = \left[P(B \mid A)P(A)\right] / P(B)$$

In this equation, *P(A)* is the **prior probability** of hypothesis *A*; in other words, it is our estimate of the probability that this hypothesis is true before we take into account new data. *P(B|A)* is the probability of the data given the hypothesis *A*; in other words, it is what we described as the **likelihood** in ML. *P(B)* is the probability of observing the data, also called the marginal probability. *P(A|B)* is the probability of *A* given *B*, or the **posterior probability**, essentially the probability that *A* is true updated for our new evidence. With BI, instead of simply accepting or rejecting hypotheses, we can calculate a degree of confidence—the posterior probability—for any given hypothesis. Note that the posterior probabilities for all hypotheses would add up to 100%.

Applied to phylogenetic analysis, each possible tree is a hypothesis (*A*) and the distribution of character data—let's say a sequence alignment for a set of taxa—is the data (*B*). The likelihood *P(B|A)* is the same calculation that we would make for ML. What really made BI so useful was the development of computational approaches that allow us to calculate posterior probabilities effectively for lots of trees. Instead of a single best tree, BI produces a set of trees with the highest posterior probabilities; typically BI reports the trees with the highest posterior probabilities that when summed equal or exceed 95%.

Model-based approaches were designed for and have primarily been used for analyzing sequence data, but there have been recent efforts to apply these methods to morphological data. Whereas modeling the evolution of changes in nucleotides is somewhat straightforward, because morphological (or other nonmolecular) characters are so diverse, the challenge is to find a model that is appropriate for that diversity of possible characters. There are currently models that have been proposed for use with morphological data that have shown promise.

5.3 Phylogeny and classification

Before we apply phylogenies to some macroevolutionary problems, we should note the relationship between phylogeny and **classification**. Classifications of organisms have existed for centuries, and the system most commonly used today was developed by Carolus Linnaeus (see *A Closer Look* below), a nonevolutionary thinker, a century before Darwin. Linnaeus, who also developed the binomial (two name) system we use for naming species, developed his system in recognition of the fact that organisms could be placed in a **nested hierarchy**, where larger groups included smaller groups, such as a kingdom containing some number of phyla. The biological reason for this pattern was, for Linnaeus, not an evolutionary one, but some reflection of the design of the Creator.

Darwin recognized that his view of multiplying species and patterns of common ancestry could provide an alternative explanation for this nested hierarchy–that the smallest, most exclusive groups contained the most closely related organisms, and larger taxonomic groups reflected more distant ancestry. Today we base our classifications on phylogenies. Typically, the taxa in a classification are meant to reflect monophyletic groups. As a result, evolutionary biologists have tended to reject taxa that do not reflect monophyletic groups, such as "reptiles," and as a result such names are used only informally (and are often printed in quotation marks), or they have been redefined: the name Reptilia is used by some workers for a monophyletic group including all traditional "reptiles" as well as birds, but other workers have argued for abandoning the name Reptilia, because it has been associated with a non-monophyletic group exclusive of birds for so many years. Figure 5.13 gives an example of how a phylogeny might be translated into a classification.

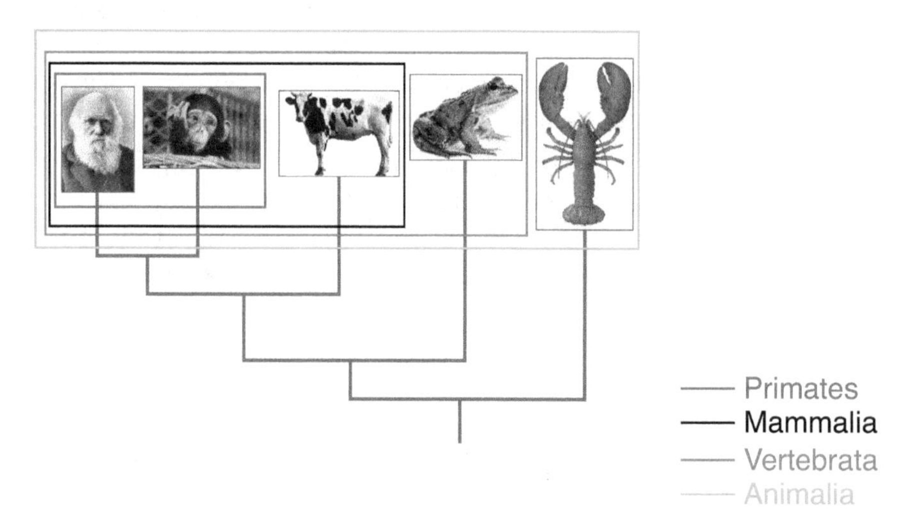

Figure 5.13. A phylogeny that has been translated into a classification based on a nested hierarchy.
Everett Historical/Shutterstock.com
apple2499/Shutterstock.com
VanderWolf Images/Shutterstock.com
Eric Isselee/Shutterstock.com
Lightspring/Shutterstock.com
© Luke Holbrook

A Closer Look: Binomial Nomenclature and Linnean Classification

Carolus Linnaeus (1707–1778; Figure 5.14) was a Swedish botanist who developed the systems for naming species and for classification that we still use today. His actual name was Karl von Linné, but we generally know him by the Latinized version of his name. His contributions have earned him the title of "father of modern taxonomy." His seminal work was *Systemae Naturae*, or "Systems of Nature," of which 13 editions were published between 1735 and 1793, the last edition coming out in multiple parts after his death. It is actually in the 10th edition of this work published in 1758 that Linnaeus introduced the systems of naming and classification.

Systemae Naturae lends its name to the field that it started, **biological systematics**. In its most general sense, systematics deals with patterns of biodiversity and the biological basis for these patterns. Today, systematics is largely concerned with phylogeny and the classifications based on them, and it explains patterns of biodiversity in terms of the Darwinian concept of common descent. **Taxonomy** is the part of systematics that is concerned with translating phylogenetic patterns into classifications.

Figure 5.14. Carolus Linnaeus, the "father" of the field of systematics.

Binomial nomenclature

Linnaeus came up with the modern scientific method of naming species, **binomial nomenclature**. Nomenclature simply refers to a system of naming things, and binomial translates to "two names," reflecting the two-part names that we use for species in scientific discourse. For instance, our own species is *Homo sapiens*.

Binomial nomenclature solves a number of problems that would exist without a standardized system and that plagued natural historians before Linnaeus. Binomial nomenclature gives every species a unique name. Before Linnaeus, and as we often do today outside of science, species were identified by common names, such as "wolf" or "robin." However, common names are not standardized, and different languages use different names for the same species. Even in English, what Americans call a moose is called an elk in Britain. There are also cases where the same common name is used for different species: "robin" refers to very different birds in America and Britain. The Linnean system avoids this by assigning a unique name to each species.

Each scientific name has two parts: the **genus** and the **trivial epithet**; our genus is *Homo*, and our trivial epithet is *sapiens* (Figure 5.15). Note that the trivial epithet is not the "species name"; the species name is both parts together (sometimes called the **binomen**). The first letter of the genus is always capitalized, and the trivial epithet is never capitalized, even if it refers to a proper noun, such as someone's name. Both parts are Latinized: that is, they are written as Latin forms, even if they are not actual Latin words, and they follow the grammatical rules of Latin. Species names are always either italicized or underlined. Because they are Latinized, they are always written in the roman alphabet, regardless of the language used elsewhere in the writing. Thus, even in a paper written using, say, Chinese characters, the species names will always appear in roman letters (i.e., the same used in English).

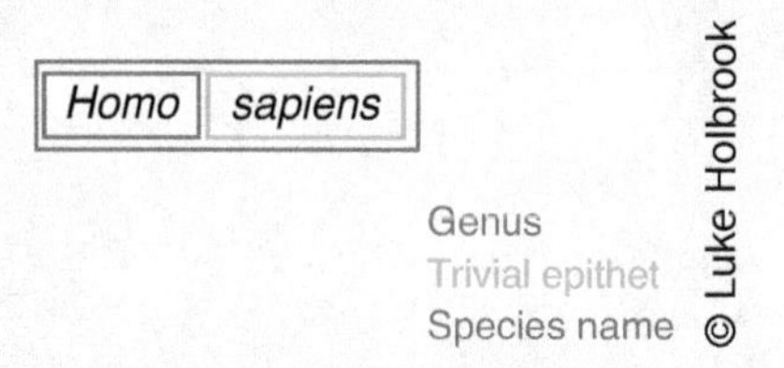

Figure 5.15. The parts of a scientific name based on binomial nomenclature.

The binomen is analogous to the two-part name familiar to most Americans, where there is a first name and a last name. In this case, the last name is meant to indicate relationship in a family, whereas the first name identifies you as a specific member of that family. We often know many unrelated people with the same first name, but we keep them straight by virtue of their different last names. The binomen works in a similar way, except that the genus is analogous to a person's last name, and the trivial epithet is analogous to a person's first name. The genus (pl. **genera**) indicates that this species is part of a specific group of related species; other species in the genus *Homo* are our closest relatives. A genus is unique to that group of species and is never repeated in another group of organisms; there is no genus of plants or insects also called *Homo*. The trivial epithet, on the other hand, can be repeated in different genera; *Lontra canadensis* is the American river otter, whereas *Castor canadensis* is the American beaver. This is why one cannot just use the trivial epithet to identify a species, but instead must use the full binomen.

Scientific names are used frequently in scientific papers, and it gets a bit redundant and tedious to write or type the same name over and over, especially if it is long. There is a convention that is used to make this easier: after the first time a scientific name is used in a paper, the genus can be abbreviated to the first letter with a period. Thus, after the first time you spell out *Australopithecus robustus* in a paper, you can thereafter refer to it as *A. robustus*.

Classification as a nested hierarchy

Systemae Naturae also gave us the nested hierarchy that comprises our classification system. A good illustration of both nested hierarchies and another type, a linear hierarchy, is the military. The ranks of officers in the army is an example of a linear hierarchy: a general ranks above a colonel, who ranks above a captain, who ranks above a sergeant, who ranks above a private.

One rank does not include any of the others: a general is not "made up" of a bunch of colonels. This hierarchy is essentially like a ladder. The units in the army, on the other hand, are a nested hierarchy: an army is made of multiple divisions, which are each made of multiple companies, which are each made of multiple platoons. Notice that in my examples of both linear and nested hierarchies that I have not included every rank that exists. The important difference to note is that, in a nested hierarchy, lower ranks are contained in higher ranks.

Linnaeus established a number of ranks to indicate which groups were more inclusive or exclusive; to put it another way, higher-ranked groups contain one or more lower-ranked groups. The species serves as the lowest rank in Linnaeus' nested hierarchy. The genus is also a rank, as it contains one or more species. Figure 5.16 illustrates the ranks in the taxonomic hierarchy. When we refer to "higher" and "lower" taxa, we are usually referring to whether they have a higher rank (more inclusive, like a kingdom) or a lower rank (more exclusive, like a genus).

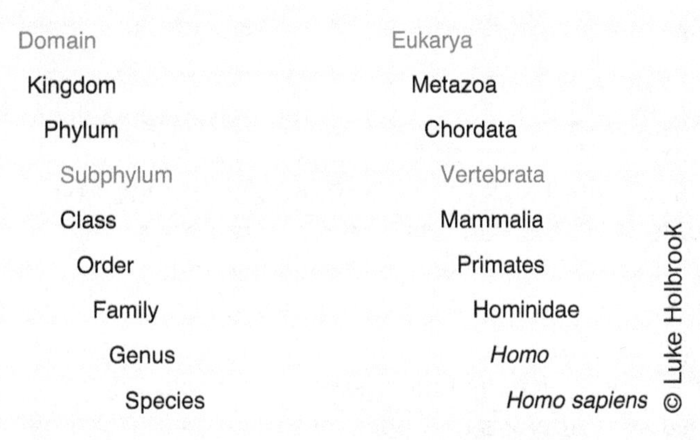

Figure 5.16. An illustration of Linnaeus's ranks for the classification of our species, *Homo sapiens*, including an illustration of a "sub"-rank (subphylum) and a new rank (domain).

Any phylogeny with enough complexity could easily exhaust the seven classic ranks of Linnaeus and still have nested groups that could be named. Linnaeus' ranks can be adapted in two ways. One is to subdivide ranks, using prefixes to indicate how they relate to the existing ranks: a subphylum is of a lower rank than a phylum but of a higher rank than a class, and a superfamily is of a higher rank than a family but a lower rank than an order. A second strategy is to simply create new ranks. The highest rank for Linneaus was the kingdom, to distinguish what he felt was the fundamental division of all life into animals and plants. Our current understanding of the diversity of life, which now includes a multitude of microbes unknown to Linnaeus, has prompted the establishment of a new rank above that of kingdom, the **domain**; we are part of the Domain Eukarya, which includes all organisms that have cells with a nucleus.

Some researchers have argued that the Linnean ranks have become uninformative: both beetles (Order Coleoptera) and mammals (Order Mammalia) are ranked as orders, but is there any real equivalency of these ranks, when there are at least one hundred times as many species of beetles as there are of mammals? These researchers have argued for eliminating Linnean ranks from classifications, but many workers still employ them. Even if distantly related taxa of the same rank are not really equivalent, the ranks do provide information within a lineage as to which groups are within which other groups.

You might be wondering who gets to assign these names of species and higher taxa. The names we use have been assigned by various researchers, and in systematics papers there is often a section on systematics that shows the different ranked taxa of interest, and each taxon is followed by the author and year of publication for when it was established, for example, Genus *Homo* Linnaeus 1758. Linnaeus himself named a great many species, so much so that often a simple "L." is used to indicate that a taxon is attributed to him.

There are some formal rules for naming anything from a species up to and including a family. A species must have a **type specimen**, a specimen that other scientists could study (and that is therefore held in some recognized scientific collection, such as a museum) that is considered to be representative of the species. A genus must have a **type species**, and a family must have a **type genus**. A family name must also always end in "-idae" for animals or "-aceae" for plants.

There are also instances (in fact quite commonly, especially in paleontology) where two or more different species have been named for specimens that are apparently part of one species. Which species name would apply? The convention is for the first name that was established to be the name that is retained. Perhaps, the most infamous case of this is that of *Brontosaurus*. *Brontosaurus* is a familiar name (and an evocative one, translating to "thunder lizard") known to most people who encountered dinosaurs as children and associated the name with an enormous sauropod dinosaur. As early as 1903, it was recognized that the specimens referred to *Brontosaurus* almost certainly belonged to the same genus as those assigned to another sauropod, *Apatosaurus*. *Apatosaurus* being the older name, the name *Brontosaurus* was largely abandoned in the scientific literature, but it continued to appear in popular works, and there was even an unsuccessful appeal to the governing body for animal taxonomy to officially use *Brontosaurus* instead of *Apatosaurus*. Interestingly, a study in 2015 of sauropod phylogeny concluded that *Brontosaurus* is actually distinct from *Apatosaurus*, though there is still some skepticism about this.

5.4 Phylogeny and prediction

Once we have a phylogeny for a group of species, we can make predictions based on the phylogeny of what else we should observe. In this section, we will explore examples of how predictions made from phylogenetic analyses accord with additional data.

5.4.1 Congruence

The characters that a particular analysis might use will often differ substantially from those of a different analysis, so we could ask how much **congruence** there is between the results of different datasets; in other words, does analysis of different sets of data for the same taxa give us similar results? When DNA sequences started to be used for phylogenetic analysis, there was great interest in how congruent results from these data would be with results from morphological data. It turns out that there is a great deal of congruence between the two. Many of the areas of conflict have turned out to be due to weak morphological support for a particular hypothesis, or a lack of rigorous studies that had analyzed the question using morphology.

Perhaps, the most interesting case of this in recent years has been the series of studies that have established the close relationship of cetaceans (whales and dolphins and kin) with hippopotamuses. Figure 5.17 illustrates two different views of the relationships among cetaceans, hippos, and other artiodactyls. Artiodactyls ("even-toed") are a group of hoofed mammals with many familiar members, besides hippos: cattle, sheep, antelope, giraffes, camels, deer, and pigs are some of the most easily recognizeable members of this group of mammals. One of the diagnostic features of artiodactyls is the peculiar shape of one of their ankle bones, the astragalus, which has two ends shaped like pulleys. (The same bone in our ankle has only a single pulley.) Living cetaceans retain only vestiges of their hind limbs and therefore don't have any ankle bones to compare with artiodactyls. Both artiodactyls and cetaceans had long been classified as separate orders of mammals (Artiodactyla and Cetacea, respectively) with different origins.

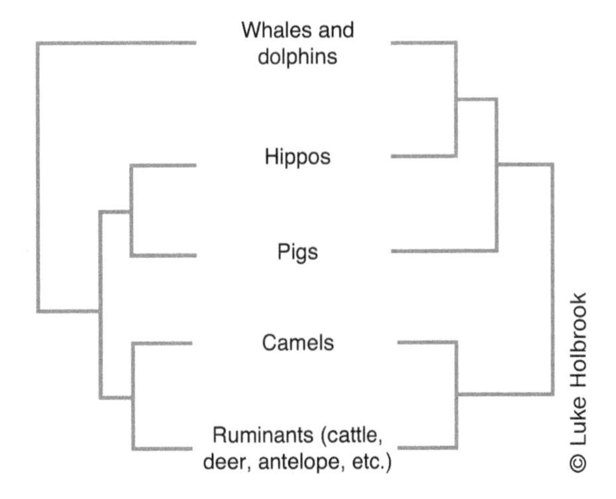

Figure 5.17. Two hypotheses of relationships among cetaceans and terrestrial artiodactyls, left based on morphology and right based on molecular data.

The origin of cetaceans has been a topic of particular interest among students of mammal evolution, since it involves the transition from a terrestrial lifestyle like that of most mammals to the fully aquatic adaptations that characterize cetaceans today. Living cetaceans have streamlined bodies, flipper-shaped forelimbs, only vestiges of hindlimbs, and nostrils that are positioned on top of the head to form a blowhole, allowing cetaceans to breathe air while just breaking the surface of the water. A number of fossil finds in recent decades had documented some early stages in the transformation of these traits, such as nostrils closer to front of the skull and more developed (though still rudimentary) hindlimbs. But these were all still very much aquatic animals, and, until the late nineties, no one had found anything that documented a cetacean that was still even somewhat terrestrial. The closest terrestrial relatives of cetaceans were hypothesized to be a group of extinct carnivorous to omnivorous hoofed mammals called mesonychids.

In 1996, *Ambulocetus*, which translates to "walking whale," was described from fossils from Pakistan in sediments around 50 million years old. *Ambulocetus* possesses characters that united it with cetaceans, as well as fully developed hindlimbs (although the ankles were largely unpreserved) and nostrils close to the tip of the snout. Superficially, it had something of a crocodile-like appearance. While there were no particular similarities to mesonychids, *Ambulocetus* gave paleontologists no reason to reject the notion of mesonychids as the closest relations of cetaceans.

Around the same time that *Ambulocetus* was described, researchers using DNA data to investigate relationships among mammals found that their data supported a close relationship between cetaceans and hippos (Figure 5.17). In other words, not only are cetaceans closely related to artiodactyls, these results said they actually are artiodactyls, because they nest within that group. This set off a series of investigations, where a number of morphologists claimed that there was stronger evidence against cetaceans belong to Artiodactyla, while subsequent molecular studies supported the hippo-cetacean relationship.

However, not all morphologists rejected the hippo-cetacean hypothesis, and subsequent discoveries brought the morphological data into greater agreement with molecular data. Some studies showed that rigorous analysis of morphological data produced results consistent with the molecular results. In 2001, two groups of paleontologists reported material from Pakistan of early cetaceans that were terrestrial and that possessed the double-pulley astragalus of artiodactyls. Thus, the results from morphological data, including data from fossils, have become more congruent with the results from molecular data.

5.4.2 Vestigial organs and character states lost

Imbedded in the story of the origin of whales is a common theme used as evidence for evolution: vestigial organs. Living cetaceans, as well as many of their fossil relatives, retain only vestiges of hindlimbs; in the case of living cetaceans, the only elements present are a poorly developed pelvis (hip bone) and a small, incompletely formed femur (thigh bone). These structures strike us as important for evolutionary explanations for two reasons. First, the structures seem to be unimportant to the way the organism currently functions; whale swimming is powered by strokes of their powerful, fluked tails, and pronounced hindlimbs would only make them less streamlined. Second, the retained structures are reduced versions of those we would expect to find in their ancestors; the terrestrial ancestors of whales would have had well-developed hindlimbs for running and walking.

Another way of thinking about vestigial organs is that their presence is a prediction given hypotheses of common ancestry. For instance, take the tree illustrated in Figure 5.18.

Given the distribution of character state X, we would reconstruct that character state as being present in the most recent common ancestor of taxa A and B, but state X would have been lost somewhere in the lineage leading to taxon B. Based on the character state distribution and the hypothesis of relationships, we can make some predictions.

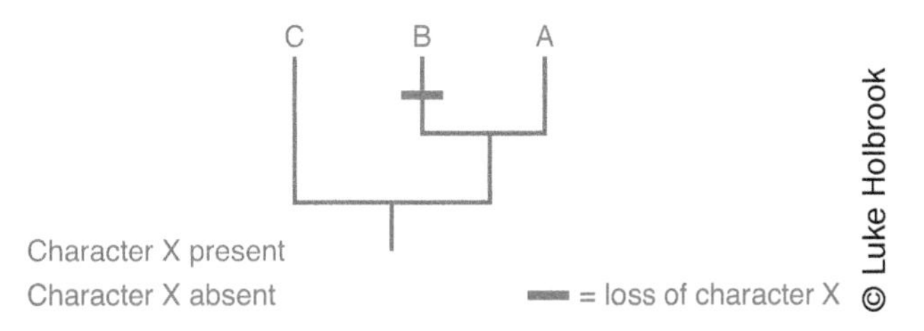

Figure 5.18. A tree illustrating how given the tree above and the distribution of character X, we would infer that character X was lost at some point in the branch leading to taxon B.

For instance, we would predict that there existed at some point taxa that possessed state X or some modification of it that were also more closely related to taxon B than to any other taxa in the tree. There are a number of examples of this sort of prediction matching what was later discovered in the fossil record, including the whale example we have discussed already. Snakes, amphisbaenians (snake-like relatives of some lizards), and caecilians (a group of mainly burrowing amphibians) all lack limbs, but all are hypothesized to be related to groups that have limbs, as illustrated in Figure 5.19. In all three cases, we now know of fossils that are early relatives of these limbless groups and that possess limbs.

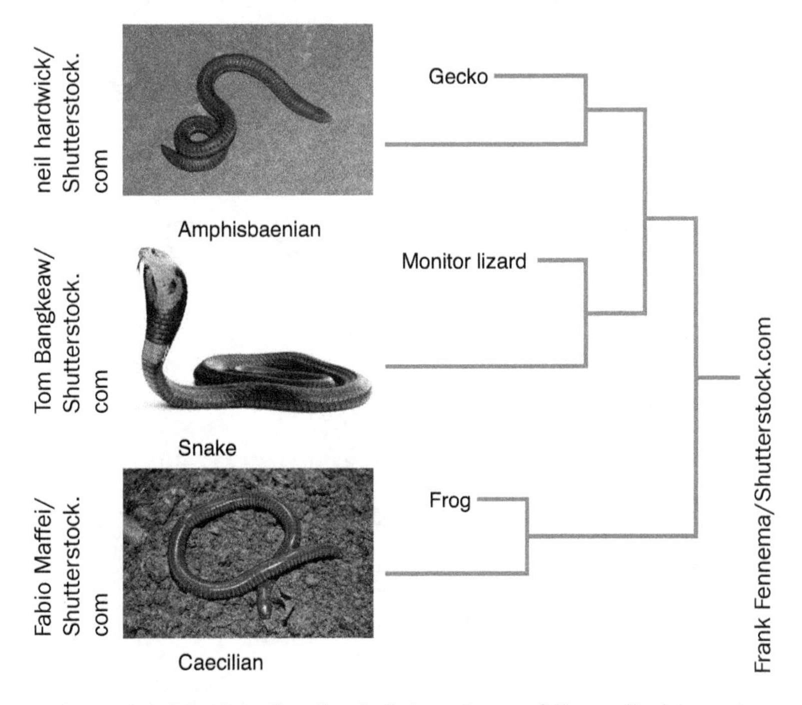

Figure 5.19. Lost and vestigial traits. On the left is a tree of three limbless terrestrial vertebrates—a caecilian, an amphisbaenian, and a snake—and one of their limbed relatives, indicating that limbs were lost independently in each lineage.

Another example comes from the two types of living jawless fishes, lampreys and hagfishes. Most vertebrates have moveable jaws, and this feature unites them in a group called gnathostomes. Lampreys and hagfishes lack jaws, which is presumably the ancestral condition for vertebrates. Hagfishes also lack vertebrae, whereas lampreys have simple vertebrae. For this reason, until recently, most morphological studies placed lampreys as more closely related to gnathostomes than they were to hagfishes, though there were some morphological studies that

Figure 5.20. A variety of mammals adapted for ant-eating. (A) A pangolin, or scaly anteater. (B) An echidna, or spiny anteater. (C) An aardwolf. (D) A tamandua, or collared anteater. (E) An aardvark.

placed the living jawless fishes in their own monophyletic group. This latter hypothesis has been supported by studies of molecular data. If lampreys and hagfishes are closely related, we would predict that hagfishes have lost vertebrae, which would have been present in the ancestors of all vertebrates. Recent embryological and genetic work (Ota et al., 2011) has demonstrated that hagfishes have vestigial vertebrae as embryos but lose them later in life, and they possess, and express to some extent, the genes involved in the development of vertebrae.

As illustrated in the case of hagfishes, another type of prediction that modern biology allows us to test is that while phenotypic states may be lost, we may still see evidence of those traits in genes of the taxa in question. For instance, several lineages of mammals have independently reduced or lost their teeth (Figure 5.20). Often, these are groups that are specialized for feeding on ants and termites, because these small but numerous insects can be collected on a sticky tongue, and teeth are unnecessary for acquiring or processing this type of food, and may even get in the way of the tongue. These mammals include a monotreme (the echidna or spiny anteater), a marsupial (the numbat), the anteaters of South America, the aardvark of Africa, the pangolins of Africa and Asia, and even a derived hyena (the aardwolf).

All of the toothless species are related in some way to mammals with teeth, so we would reconstruct the possession of teeth as ancestral for them and later lost in these lineages. We might further predict that these toothless taxa would still retain the genes that specifically govern tooth development. In mammals that need and use teeth, there would be strong selection against individuals with mutations in these genes. In these toothless mammals, there would be no such selection, since mutations in these genes would not affect their fitness. Therefore, we would expect to find more mutations in these genes in toothless forms, and these genes might ultimately become nonfunctional pseudogenes in toothless taxa. This is in fact what researchers found for the tooth development genes they studied.

5.4.3 Congruence with the fossil record and geology

Besides congruence between phylogenies produced from different datasets, we can make predictions about how phylogenies will be congruent with other types of information on earth history. For instance, how do the patterns that we recover from phylogenetic analysis match what we see in the fossil record, or with the patterns of continental drift inferred from plate tectonics?

If we compare phylogenies to the fossil record, we would expect that the timing of appearance of various groups should correspond to the sequence of nodes on the tree. The range of time over which a taxon is known in the fossil record is often represented in a diagram reflecting **stratigraphy**, the relations of various rock units in time. From these relations (using the dating methods discussed in Chapter 2), we can reconstruct which fossils found in the rocks are older or younger than others. The information is summarized in a diagram like the one in Figure 5.21 indicating with a vertical line the extent of time that a taxon has been around, from its first appearance in the record until the present—or until its extinction.

The correspondence between the fossil record and phylogeny is remarkably good, despite the fact that the fossil record is extremely incomplete. Fossilization is a rare event requiring unusual conditions, and thus less than 1% of the diversity of past life is estimated to have been preserved, much less discovered by palaeontologists, and even that small percentage

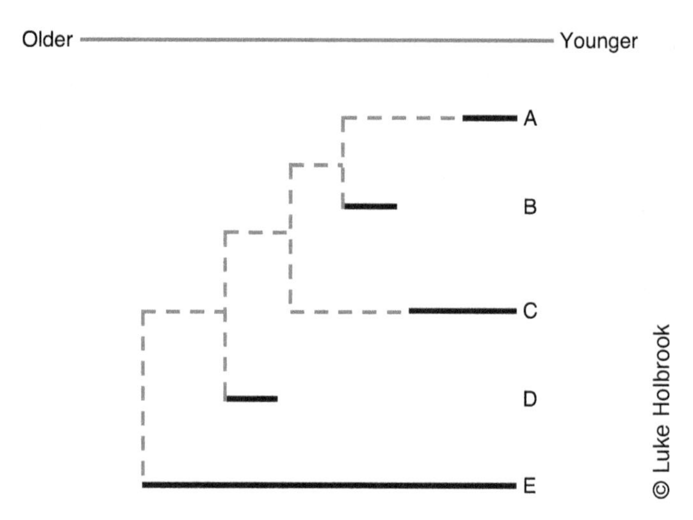

Figure 5.21. A hypothetical phylogeny with stratigraphic ranges of taxa indicated as solid lines and "ghost lineages" indicated as dashed lines.

is biased toward taxa with hard parts that are more easily preserved and that lived in or near environments with appropriate conditions for fossilization. Note that the analyses that produced the phylogenies used for these comparisons did not use data on stratigraphic appearance of the taxa analyzed to determine the phylogeny. In other words, the study of the age of the fossils was done independently of the phylogenetic analysis, so the correspondence of the two is independent support of Darwin's concept of common ancestry, and not a product of circular reasoning.

5.4.4 Molecular clocks

The analysis of molecular data for phylogeny has also provided an alternative way of studying the timing of evolutionary events. We expect to see—and do see—greater differences when comparing the genes of more distantly related taxa. Therefore, we would expect the number of differences to be correlated with the amount of time that has passed since two taxa diverged from their common ancestor. A **molecular clock** attempts to measure this amount of time by establishing the rate of mutations that occur along a branch. This often requires some data from fossils to calibrate the rate of genetic change, but there are cases involving the relatively rapid evolution of viruses where the rates can be estimated directly from our knowledge of the virus. Figure 5.22 gives an example of a dated tree for vertebrates based on a molecular clock.

One of the more remarkable instances where a molecular clock was applied involved a case where over 400 children became infected with human immunodeficiency virus (HIV), the virus that causes acquired immune deficiency syndrome (AIDS), in a hospital in Benghazi, Libya, in 1998. The tragedy was deemed by the Libyan government to be a crime of terrorism perpetrated by a Palestinian doctor and five Bulgarian nurses who had arrived at the hospital that year. Despite evidence of their innocence, they were still convicted and sentenced to

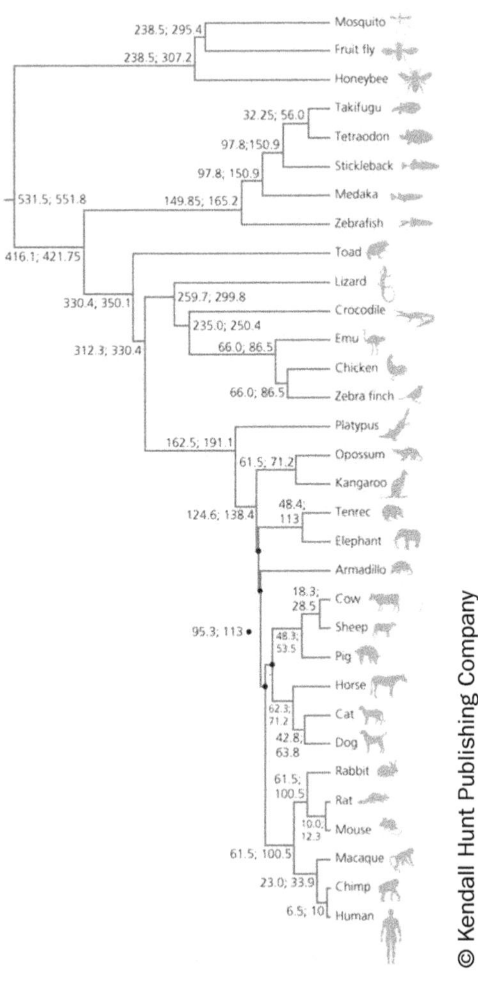

Figure 5.22. Dated tree of vertebrates (and some invertebrates) based on a molecular clock. Numbers at nodes indicate how long ago that divergence occurred in millions of years.

death in 2004, after having already spent nearly six years in jail. One study by Oliveira et al. (2006) that helped to establish their innocence was a 2006 molecular clock analysis of HIV samples obtained from 44 of the infected children. The researchers were able to establish that the age of the most recent common ancestor for the HIV samples from the children predated the arrival of the six accused medical staff. Ultimately, political pressure resulted in the accused being released to Bulgaria, where their sentences were commuted.

5.5 Convergence

Darwin noted that his evolutionary ideas could explain some interesting patterns among species. Part of the geographic evidence that he mustered in support of his ideas was that although very similar kinds of habitats occur in different parts of the world, they are not inhabited by the same kinds of organisms; polar bears are adapted to cold climes, but they occur only in the highest latitudes of the Northern Hemisphere, and not in the similar latitudes of Antarctica (contrary to the depictions of polar bears and penguins in some advertisements). Even more interesting from an evolutionary perspective, similar habitats in distantly separated places are sometimes inhabited by distantly related organisms with similar adaptations. This is a form of homoplasy known as **convergence**, where lineages independently converge upon similar "solutions" due to similar selective pressures.

For instance, beavers are found in the Northern Hemisphere, where they are adapted to the familiar beaver life of living in ponds, eating vegetation, and constructing lodges of branches. Some other species of rodents have similar habits and adaptations for a semi-aquatic life. The nutria or coypu is a South American beaver-like rodent (Figure 5.23). It was introduced to North America for ill-advised reasons, where it has caused a great deal of ecological and economic damage. Beavers and nutrias have similar ecological niches, but in other features, particularly the anatomy of their skulls, the two are different, and nutrias are more similar in these respects to certain other native South American rodents, like porcupines. Darwin explained this as two species that don't share a particularly close common ancestry, but on whom natural selection has had a similar effect. If this is true, then a phylogeny of rodents should reflect this.

Brian Lasenby/Shutterstock.com

Zmrzlinar/Shutterstock.com

Figure 5.23. Convergence between beavers and nutrias. The animal on the left is an American beaver (*Castor canadensis*). The animal on the right is a nutria (*Myocastor coypus*). Both species have similar adaptations for a semiaquatic life and build dams and lodges in freshwater environments. Both are also rodents, but they belong to different families and are each more closely related to certain other, nonbeaver-like rodents than they are to each other.

Modern studies of relationships among rodents support the hypothesis of convergence, as phylogenetic analyses both of morphology and of genetic data clearly support the hypothesis that nutrias are more closely related to other South American rodents than they are to beavers. Thus, the features that nutrias and beavers share that are related to the similarities in their lifestyle must have evolved independently.

There are many other examples of convergence (Figure 5.24). We have already observed that a number of mammal lineages have evolved forms with long snouts and reduced or absent teeth specialized for eating ants and termites. The similar special societies that characterize ant and termite colonies evolved spearately in these different orders of insects. There are multiple lineages of mammals, including multiple families of rodents and marsupials, that have evolved gliding membranes. Desert environments in different parts of the world are inhabited by similar but unrelated types of plants with spines and fleshy stems.

Figure 5.24. Convergent evolution illustrated by similar adaptations seen in certain marsupials and placental mammals. A sugar glider (*Petaurus breviceps*; A) has similar gliding adaptations to those of a southern flying squirrel (*Glaucomys volans*). The numbat (*Myrmecobius fasciatus*) is a marsupial anteater with adaptations similar to those of placental anteaters, like the tamandua (*Tamandua tetradactyla*).

Marsupials, particularly in Australia, include a number of species that share striking similarities with placental mammals on other continents (Figure 5.24). Sugar gliders (Genus *Petaurus*) are marsupials that closely resemble flying squirrels (Genus *Glaucomys*) in the possession of gliding membranes extending between their wrists and ankles, among other things. There are marsupial anteaters (*Myrmecobius fasciatus*) and marsupial moles (*Notoryctes typhlops*) that resemble placental anteaters and moles. The extinct thylacine (*Thylacinus cynocephalus*; sometimes known as the Tasmanian wolf or Tasmanian tiger), the last individual of which died in a zoo in 1936, was a wolf-like carnivore. Perhaps, the most striking example comes from the fossil record: the fossil marsupial *Thylacosmilus* resembled a sabre-toothed cat, including long canine teeth and flanges on the lower jaw that protected them.

5.6 Other predictions of macroevolution

There are other phenomena that support the notion of common ancestry among organisms. Darwin hedged on whether all life had a single ancestor, but the hypothesis of a single common ancestor for all life has been supported. One form of support is the commonalities of structure and function of cells across all major divisions of life. All living things are made of cells (at least one!), have cell membranes, have DNA that goes through replication, transcription, and translation, and have many similarities in their metabolic pathways.

But what if all of those commonalities are shared not because they were inherited from a common ancestor, but because they are simply the best, or even the only way to build a cell and have it function? The strongest evidence for common ancestry, therefore, would be traits that are not only shared by all living things, but that represent only one of multiple ways that living things could perform the functions of these traits. An analogy would be languages. Languages have an obvious function (communication), and they are inherited from generation to generation through cultural mechanisms. However, while it is useful to have a language, that does not mean that all cultures need to have the same language. Thus, societies and cultures that share a common ancestry often share a language (or language family) that is different from that of other societies.

There is a language that is common to all life: the **genetic code**, discussed earlier in this chapter. The codons that correspond to specific amino acids are the same for all living things. While it is important for an organism to have a genetic code, there is no functional reason why every species must have the same code, unless it is due to inheritance.

A similar argument applies to the chirality of amino acids. The arrangement of atoms on an amino acid can be "left handed" or "right handed," what is referred to as the **chirality** of the amino acid. For proper biochemical function, all amino acids in an organism must be of the same chirality, but it does not matter whether they are all left or right handed. In fact, the chirality of all amino acids in living things is left handed. The fact that none of them is right handed suggests that left-handed chirality was present in the ancestor of all life and has been inherited in all of its descendants. If living organisms today were actually descended from multiple origins of life, we would expect that at least some of them might show right-handed chirality.

Literature Cited

Oliveira, T. de, O. G. Pybus, A. Rambaut, M. Salemi, S. Cassol, M. Ciccozzi, G. Rezza, G. Castelli Gattinara, R. D'Arrigo, M. Amicosante, L. Perrin, V. Colizzi, C. F. Perno, and Benghazi Study Group. 2006. "HIV-1 and HCV Sequences from Libyan Outbreak." *Nature* 444:836–37.

Ota, K G., S. Fujimoto, Y. Oisi, and S. Kuratani. 2011. "Identification of Vertebra-Like Elements and Their Possible Differentiation from Sclerotomes in the Hagfish." *Nature Communications* 2:373.

CHAPTER 6

The Predictable Consequences of Speciation and Biogeography

The insight that was perhaps most important for driving Darwin down the path of evolution was that species on different islands descended from a common ancestor that multiplied into several descendant species. The specific incident that inspired him was the realization that each Galápagos island had its own species of mockingbird, and yet these species were very similar to each other (and quite different from the species on the mainland). Thus, the observations on the mockingbirds led to his theory of multiplication of species as a consequence of evolution. One thing Darwin did not elaborate on was exactly how those species formed from that one ancestral species, a fact that some have found curious. In fact, this isn't that surprising when you consider that Darwin had to first convince his audience that species could change at all. In this chapter, we'll explore some of the ideas that have been developed since Darwin to explain the process of multiplication of species by evolutionary processes.

6.1 Species

We can't really have a discussion of multiplication of species without some sense of what **species** are. In the most general sense, species are kinds of organisms that can be distinguished from one another. When members of a species reproduce, they give rise to more members of the same species and not to members of other species. Thus, although male and female humans can be distinguished from one another, we don't consider them to be different species, because their offspring consist of more male and female humans. On the other hand, we recognize humans as separate species from chipmunks, ants, and oaks, because each of these kinds of organisms is distinct from the others, and each gives rise to more of its own kind and only its own kind.

This very general definition of species is what naturalists recognized well before Darwin. What needed explanation, particularly after the advent of evolutionary biology, was the biological reasons for why species remain distinct. As it turns out, we can explain this at multiple levels, and thus we have multiple **species concepts** that are used in biology. We will focus on two concepts here.

At the end of Chapter 4, we discussed how migration tends to keep populations genetically similar by maintaining **gene flow**, essentially the ability of alleles to be exchanged between populations through interbreeding of individuals from different populations. Disruption of gene flow allows two populations of the same species to evolve in separate directions, and therefore become distinct. Thus, if gene flow is disrupted between two populations, what we will observe is two populations that are distinct from one another and that give rise to more individuals like themselves. The absence of gene flow alone could thus be used to distinguish two species. This is essentially the basis of the **phylogenetic species concept**. In order to establish that two populations are distinct species by this concept, one needs to establish that there is no gene flow between their populations.

Activity

Imagine you have two populations that you think might be separate species. You can collect data on alleles for, say, twenty different genes from individuals from both populations.

1. *Articulate the alternate and null hypotheses regarding the number of species represented by these populations.*
2. *If the alternate hypothesis is correct, what would you expect to find in terms of a comparison of allele frequencies for the two populations? Would they have the same frequencies for all genes? Would they have different frequencies for all genes? Would some genes have the same frequencies and others have different frequencies?*
3. *Compared to your answers for #2, what would you predict if the null hypothesis is true?*

*Recently, researchers have proposed that African elephants include two species, savannah elephants (*Loxodonta africana*) and forest elephants (*Loxodonta cyclotis*). These two types of elephant had been recognized as having certain morphological differences, but until this recent work they were considered to be members of the same species. One study collected sequence data from individuals of savannah and forest elephants, as well as some specimens that could not be assigned to either.*

4. *If you made a phylogenetic tree from these sequences, what would you expect it to look like if the two types of African elephant were separate species? How would the tree look if they were the same species?*
5. *Look at the tree in Figure 6.1. How does this compare to the predictions you made above? Does this support considering these elephants as one species or as two?*

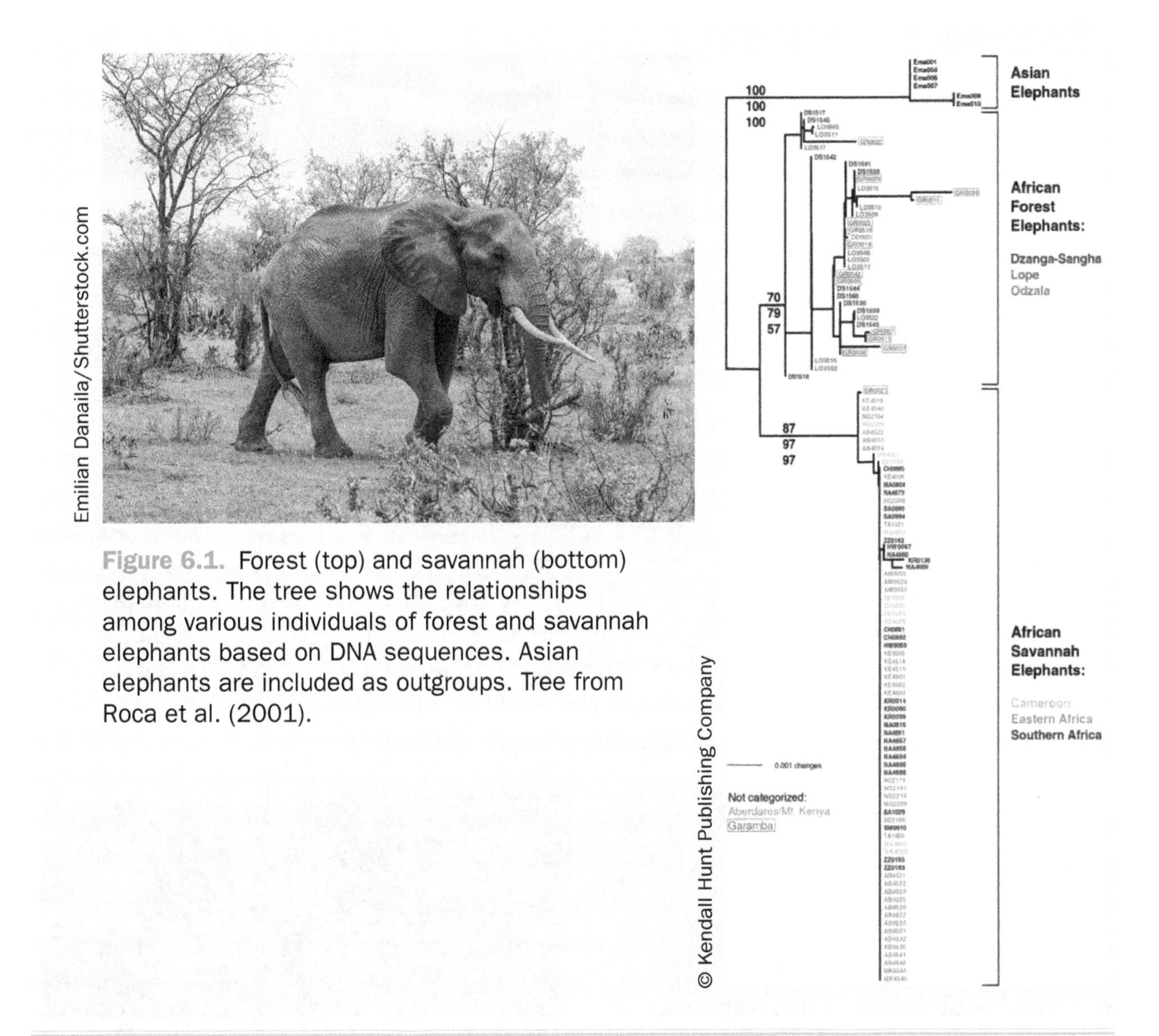

Figure 6.1. Forest (top) and savannah (bottom) elephants. The tree shows the relationships among various individuals of forest and savannah elephants based on DNA sequences. Asian elephants are included as outgroups. Tree from Roca et al. (2001).

Another way of thinking of the phylogenetic species concept is that, if two populations have been evolving separately (i.e., without gene flow between them), then they are separate species. One reason we might observe such a case is when two populations are geographically isolated, or **allopatric**, perhaps because they are on different islands (or different island-like situations, like isolated mountaintops or unconnected ponds), or because some geographic barrier has separated them (such as the rising of a mountain chain, the emergence of a river, or even the separation of land masses through continental drift). In these cases, if we theoretically took individuals from the different isolated locations and gave them the opportunity to mate, they might very well produce fertile offspring. In other words, the two populations might not yet have evolved differences sufficient to prevent gene flow if they were in the same place (or **sympatric**). The presence of biological, rather than simply geographic, barriers to gene flow is referred to as **reproductive isolation**. While the phylogenetic species concept does not require reproductive isolation, another species concept does.

The **biological species concept** identifies a set of populations as comprising a species if they can have gene flow among each other but not with any other population. Therefore, a biological species is a set of populations that are reproductively isolated from every other population but not from the constituent populations of the species. Thus, reproductive isolation is required for identifying species according to this concept. Reproductive isolation includes a variety of biological (rather than simply geographic) reasons why populations might fail to exchange genes; in other words, **reproductive isolating mechanisms** are based on how the organisms live, not simply where they live. These could be things that prevent individuals of different species from ever successfully producing a fertilized egg: different species may have different mating seasons, different habitats in which they dwell, different behaviors to elicit mating, or incompatible reproductive anatomy. Such isolating mechanisms are called **prezygotic**, because they prevent the formation of a fertilized egg, or **zygote**. In some cases, members of different species can actually mate and produce a fertilized egg, but they are still reproductively isolated. In such cases, the resulting hybrid offspring ultimately fail to breed with members of the parent populations, either because they don't survive or they are simply unable to reproduce. In some cases, hybrid offspring simply fail to survive, perhaps as early as the zygote stage but even as late as just prior to sexual maturity. In other cases, the hybrids are not as fertile as their parents, the classic case being the sterile mule produced by the interbreeding of horses and donkeys. In these cases, hybrids are effectively selected against, and the result is reproductive isolation, by what is called a **postzygotic** mechanism, keeping the species separate (Figure 6.2).

Figure 6.2. Fox squirrels (*Sciurus niger*; left) and gray squirrels (*Sciurus carolinensis*; right) have overlapping ranges but do not interbreed because they have different breeding seasons. This is a kind of prezygotic isolation. Mules are hybrids of domestic horses (*Equus caballus*) and donkeys (*Equus asinus*). Mules are an example of postzygotic isolation of these two species, because, while individual mules can survive quite well, they cannot interbreed with each other or with members of their parent species.

The biological species concept is the older and better established of the two species concepts, but it is more difficult to apply, since it is not always possible to test whether individuals from two populations are truly reproductively isolated. Keep in mind that only one form of reproductive isolation needs to be in effect for two populations to be separate species, so one might need to test all of the different kinds of isolating mechanisms before feeling confident that two populations are not reproductively isolated from each other. Thus, the phylogenetic species concept is often the one that is applied in many studies, because it is easier to establish the presence or absence of gene flow than the presence or absence of reproductive isolation. Nevertheless, when we talk about the formation of new species, or **speciation**, we are generally interested in how reproductive isolation evolves, since many species clearly exhibit no gene flow despite being sympatric.

The emphasis on gene flow is important when we apply the biological species concept. As long as there can be gene flow between two populations, then they are considered to be the same species. In some cases, a species may include populations that exhibit reproductive isolation, but gene flow is still possible because intervening populations are not reproductively isolated from the apparently reproductively isolated populations. For instance, consider a species that has three populations—north, south, and one in between—where the north and south populations are so different that they would be reproductively isolated even if sympatric. However, as long as both the north and the south populations can interbreed with individuals in the middle population, gene flow will be possible from the north to the south, and vice versa. Thus, the apparent reproductive isolation of the north and south populations does not automatically make them separate species.

Gene flow between intervening populations can lead to some unusual situations when delineating species. One example is **ring species**. In this case, two sympatric populations may appear to be reproductively isolated and therefore separate species. However, in a ring species, other populations linking the two reproductively isolated populations are arranged geographically, so that gene flow is possible between the reproductively isolated populations through interbreeding with the connecting populations. A number of examples of ring species have been documented, including in salamanders and birds.

Now that we have a sense of what it means to be a distinct species, let's look at models of how an ancestral species can give rise to two (or more) descendant species.

6.2 Allopatric speciation

The Origin of Species was Darwin's device for arguing that species originate as descendants of other species, but he never fully developed the mechanism by which this could occur. That goal was achieved in the twentieth century with the theory of **allopatric speciation**. As we learned in the previous section, allopatric refers to some sort of geographic separation between things, so allopatric speciation requires some sort of geographic separation (or something equivalent) to initiate this process.

The process of allopatric speciation is outlined in Figure 6.3. We start with a population that then becomes divided by a geographic barrier to gene flow. Such a barrier could be a physical change in the landscape, like the rising of a mountain chain, or a river cutting through a region, or even something like the Isthmus of Panama dividing the Pacific Ocean from the Caribbean Sea. The geographic change could be the product of climate change, such as rising sea levels submerging part of an island to create two islands of higher ground, or formerly widespread cold climes during an Ice Age retreating to the mountaintops with the return of warmer mean annual temperatures. We could also get the same effect from some individuals being displaced to an isolated location. For instance, a few individual flying insects or plant seeds might be swept by an unusual wind from one island to another, or some larvae of marine invertebrates might drift from one seamount to another. In any case, the result of this separation is that gene flow no longer connects the two populations.

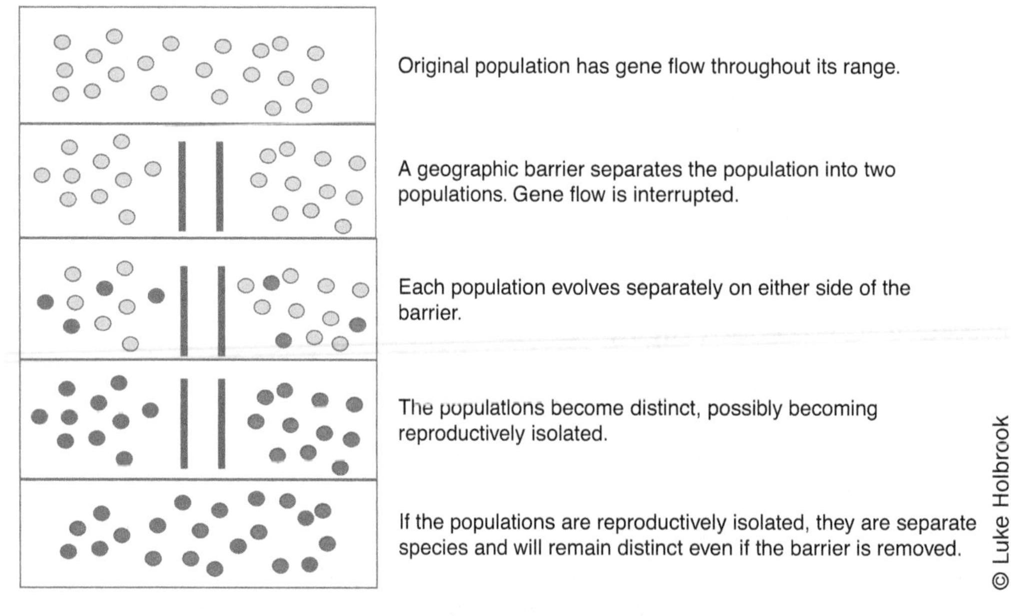

Figure 6.3. An illustration of the mechanism of allopatric speciation.

Over time, evolutionary forces of some sort may act on one or both populations; migration would not be one of them, at least not migration between the two populations. But natural selection could act on both populations. If the environments of the two geographic locations are different, then selection will act on these populations differently. Mutation could result in alleles appearing in one population that don't exist in the other; in that case, with different alleles on which it is acting, selection might produce different results even if the environments are similar. Finally, genetic drift can affect one or both populations, particularly if at least one population is small. In the case of a few individuals founding a new population on an island, the effect of genetic drift can be pronounced. Even if drift is only strongly affecting the new population, the results will be two very (genetically) different populations located in different places.

Being different does not, however, constitute being different species, at least not biological species. If the geographic barrier could be overcome, it might still be the case that the individuals in the populations could interbreed, and gene flow would be restored, and the populations would still be part of the same species. If the interglacial period ends and cold extends back down the mountains so that the Ice Age refugees on the mountaintops can now spread between the mountains, will those individuals interbreed and produce fertile offspring, or not? Ultimately, the answer to this depends on whether the differences that have evolved in the separate populations result in reproductive isolation.

If two formerly geographically isolated populations expand their ranges to become sympatric, at least in some part of their ranges, and there is no gene flow between them, then they must be reproductively isolated, and therefore we would consider them to be separate species. Note that such populations would fulfill both the phylogenetic and biological species concepts. Note also that, if this is indeed a natural pattern that is responsible for multiplication of species, then we can make predictions about the relationship between species and geography.

6.3 Speciation in space and time

The pattern Darwin saw with the Galápagos mockingbirds is one we might predict from the hypothesis of multiplication of species and allopatric speciation, namely that there would be predictable relationships between related species and geography. The island mockingbirds were more similar to each other than they were to their ancestors on the mainland, and we could extend that further by asking how the relationships among the mockingbirds relate to the manner that we predict they came to populate the islands. For instance, we might expect the islands closest to the mainland to have been populated first and those furthest from the mainland to be populated last, so birds on far islands should be more closely related to each other than they are to birds on near islands. Or, if the birds came to the islands as they formed from volcanoes under the sea, we might expect the production of new species to follow the geological formation of the islands; so, birds on the youngest islands would be most closely related, and their next closest relatives would come from the next youngest islands. We can look for similar patterns with other species on other islands.

One of the best examples of island speciation is the radiation of fruit flies on the islands of Hawaii. Over 800 species of fruit flies are found on these islands. Figure 6.4 illustrates how the Hawaiian islands are arranged. Oceanic islands, such as those of the Galápagos or Hawaii, typically form as the tops of undersea volcanoes that have risen above the surf. Such islands often form in a region of the ocean prone to volcanism, and thus volcanic islands will form sequentially over time. The island chain of Hawaii was formed by occasional bursts of magma emerging from the ocean as the Pacific plate moved over a "hot spot" prone to such eruptions. Because the islands form in a sequence over time, species can't get to a particular island until it has formed. Thus, we might expect the sequence of colonization by new species to start with a migrant from the distant mainland, which evolves to become a new island species. When a new island appears nearby, colonists from the first island species will eventually get there and spawn a new species. As new islands arise, the new species that arise there should be derived from, and therefore most closely related to, the species on the nearest older islands.

Figure 6.4. The islands of Hawaii. Note that the older islands are on the left and the younger are on the right.

Activity

1. *Articulate the alternate and null hypotheses based on the question of whether allopatric speciation is responsible for speciation of the Hawaiian fruit flies.*
2. *What do you predict should be the relationship between the phylogeny of the fruit flies and the geological formation of the islands, based on the alternate hypothesis?*
3. *What do you predict should be the relationship between the phylogeny of the fruit flies and the geological formation of the islands, based on the null hypothesis?*
4. *Compare your predictions to Figure 6.5. How do the results compare to the predictions?*

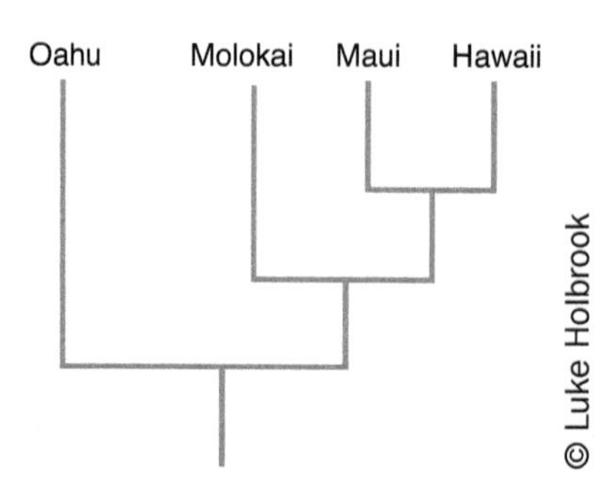

Figure 6.5. A phylogeny showing the relationship between flies from different Hawaiian islands. Note the relationship between the phylogeny of flies from different islands and the ages of the islands.

If we look at the relationships among the flies, they match the formation of the islands in the manner we would predict consistent with the hypothesis of allopatric speciation. The species on the youngest islands are most closely related to those on the next youngest island, and species on older islands are more distantly related as the islands increase in age of formation.

6.4 Disjunct distributions and their predictions

There are many cases where closely related species are geographically close to each other, which is something we would expect as a result of allopatric speciation. But what of cases where closely related species are not geographically close to each other? Do these cases refute multiplication of species?

Consider the following example. The genus *Alligator* has two living species, the American alligator (*Alligator mississippiensis*) and the Chinese alligator (*Alligator sinensis*). The current distributions of these species are **disjunct**; in other words, they are geographically separated, and by a wide margin. Once endangered, the American alligator is found today throughout the southeastern United States. The Chinese alligator is extremely endangered and is restricted to a small area of eastern China; historically, it had a range throughout much of China. Even considering the larger historical ranges of these species prior to humans affecting their distribution, American and Chinese alligators have never been geographically close to each other in the wild. Allopatric speciation, though, predicts that closely related species should be geographically close.

Arto Hakola/Shutterstock.com

reptiles4all/Shutterstock.com

Figure 6.6. The American alligator (*Alligator mississippiensis*; left) and the Chinese alligator (*Alligator sinensis*; right).

Brainstorm

Think of possible explanations for this disjunct distribution and discuss them with your classmates.

Of course, the explanation evolutionary biologists would offer for this disjunct distribution is that these species were once geographically close, if not to each other then to related species that fill the gap in their distribution, but they have become separated over time. This could be due to geological processes like continental drift, or because the populations that connected them have become extinct. Based on these explanations, we can make testable predictions.

Activity

Predict what we would expect to find in the fossil record if we explain the disjunct distribution of alligators as a relic of a more widespread distribution of alligator species.

In fact, the fossil record of the genus *Alligator* includes species distributed across Eurasia and North America. This record includes several extinct species, and it suggests that the current range of *Alligator* species is only a fraction of what it used to be. Why should this be? Alligators today inhabit warm, wet environments, and this was likely the case for their extinct relatives. Such warm habitats were more extensive in the past, especially when the genus *Alligator* first appeared, about 37 million years ago. Other members of the alligator family (Alligatoridae) have even been found in the Arctic Circle from sediments 52 million years old, during an even warmer time in Earth history. More recently, in the past few million years, not only have temperatures continued to cool down from this "greenhouse" world of the past, but periods of glaciation, or Ice Ages, have pushed cold climates southwards for tens of thousands of years. As a result, the habitats suitable for creatures like alligators have become more restricted and fragmented, such that eventually only the two species that we know today survived to represent this genus.

6.5 Niches and the ecology of species

When we find species living sympatrically, we often notice that not only do they look different from each other, they also "do" different things; in other words, we don't often see two species in the same environment in the same place living in exactly the same way. It is almost as if the species in an area have divided up the resources and the ways of making a living such that each species fills one unique role. Why is this the case? To answer this question, we need to understand how species compete for resources.

When individuals in two different species use, and therefore compete for, the same resources, this is called **interspecific competition**. Note that this is fundamentally different from the kinds of competition we discussed in Chapter 3, which was between individuals of the same species (**intraspecific competition**). To help us grasp the effects of competition between species, we will introduce the concept of a **niche**. The term **niche** is used to describe the set of resources and circumstances that individuals in a species exploit in their pursuit of survival and reproduction. This could include the foods individuals eat, the habitats they use, the hiding places they need, their sources of water, the types of soil they need, and so on.

The possible resources and circumstances that a species might use are often quite varied. The entirety of the possible niche that a species could fill is called its **fundamental niche**. In fact, in a given environment, individuals in a species typically only do some of the things that comprise their fundamental niche, because that environment does not support everything in the fundamental niche; there may be certain foods that the species can eat but that are not present in that location. What the species actually does is called its **realized niche**. Note that the realized niche should be some part of the fundamental niche.

What if two species living in the same place have overlapping fundamental niches? In other words, what if they are in competition for certain resources? For example, imagine two species of fruit flies, where one species eats bananas and mangos, and the other eats bananas and breadfruit. In an environment with bananas, individuals from both species will be competing for bananas. There are two effects of competition to consider. First, in the competition between banana-eating individuals of both species, it is possible, if not likely, that one species will be better at exploiting that resource than the other. Also, if other individuals in a given species can survive without eating bananas, for example, by eating a different fruit in that area, they will have an advantage over the individuals competing for bananas.

If we think about the consequences of this, we would expect selection against individuals that eat bananas and are worse at exploiting that resource than banana-eating individuals of the other species. As a result, eventually we would expect that the only species eating bananas would be the one that is better at exploiting bananas, and the only individuals of the other species that would be present are those that eat something else. In essence, one species evolved to no longer be in competition with the second species, by changing its realized niche.

But what if there were no fruits other than bananas available? What if the only niche that either species could realize was one of eating bananas? Because the effective fundamental niches of these two species completely overlap in this area, the species that is worse at exploiting bananas would not be able to evolve to avoid competition. The result would be the extinction of the species that was less competitive.

What we have just described is the basis for the **competitive exclusion principle**, which essentially says that two species cannot remain in competition for a limiting resource; either one species must evolve to occupy a different niche, or one species must go extinct. As a result, two species living in the same area never occupy the same niche.

The origins of the competitive exclusion principle are ascribed to the work of Georgii Gause (and the principle is sometimes referred to as Gause's Law). Gause demonstrated competitive exclusion in cultures of the microscopic one-celled organism *Paramecium*. He grew populations of three species of *Paramecium* in cultures of yeast and oatmeal. In some cultures, he grew only one species in isolation, while in others he included members of two species. He assessed the change in the populations over time, and the results are depicted in Figure 6.7.

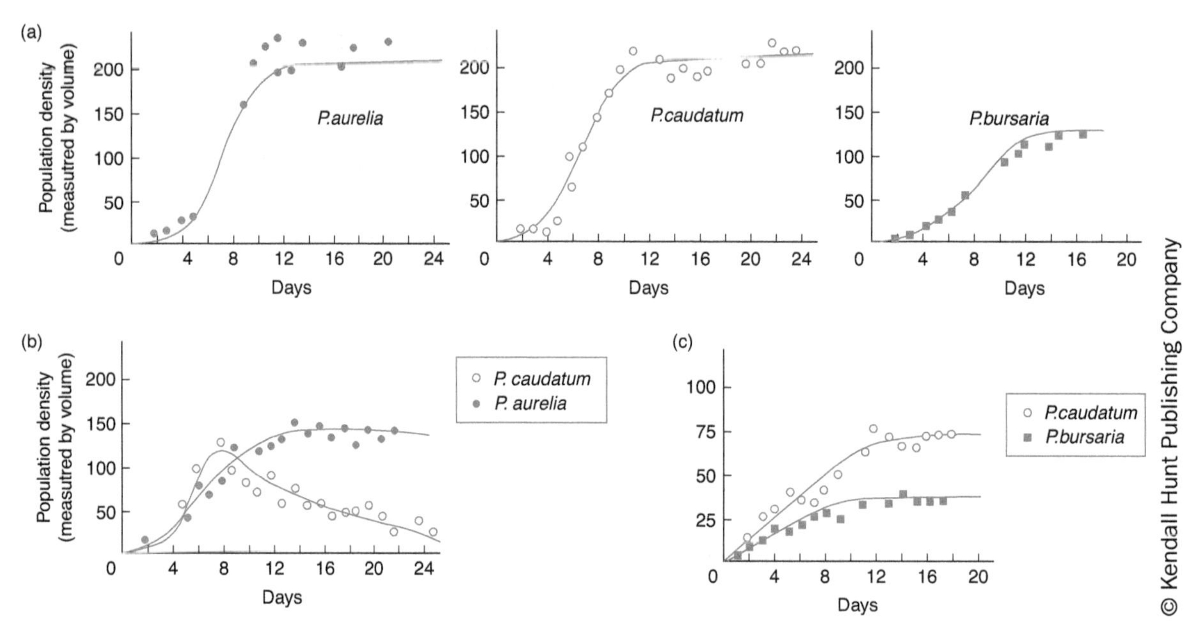

Figure 6.7. Gause's experiments. The top set of graphs shows the growth of populations of different species of *Paramecium* cultured in isolation. The bottom graphs show population growth for different species when they were cultured together.

The species grown in isolation exhibited the characteristic logistic population growth that we discussed in Chapter 3. When two species were present in a culture, however, two things could be observed. First, the population growth of both species was diminished compared to how they grew in isolation. Second, in some cases, one species would be driven to extinction.

Gause's experiments demonstrate a type of interspecific competition called **exploitative competition**. The differences in the population growths of the two species cultured together was a reflection of one species being better able to exploit the resource than the other. Thus, the effect of the better-competing species on the worse-competing species was indirect and a consequence of the better competitor simply being more effective at getting the resource. Interspecific competition can also take the form of **interference competition**, where one species directly affects the other species with which it is competing.

An example of interference competition comes from a study of barnacles growing in the intertidal zone on the coast of Scotland by Connell (1961). Barnacles are crustaceans like crabs and lobsters, but they lead a very different lifestyle from those mobile species. Barnacles are filter feeders and attach their conical shells, which look a bit like stubby volcanoes, to rocks and other surfaces. They stick their feet out of their shells and filter particles with their feathery appendages. They can survive out of the water by closing up the top of their shells. Barnacles grow by adding to the base of their shells, thus increasing both the height and the width of the cone.

The two species in this study, *Chthamalus stellatus* and *Balanus balanoides*, differ in how high they are found in the intertidal zone, the area of the coast that is exposed at low tide but submerged at high tide. *Chthamalus* is found higher up, closer to the high tide line, whereas

Balanus is found closer to the low tide line. When *Chthamalus* produces larva, many of them drift down to the *Balanus* zone, so why do they not establish themselves there?

The answer is competition, and specifically interference competition. When *Chthamalus* larvae settle in a spot between *Balanus* individuals, the growth of the *Balanus* eventually literally squeezes out the *Chthamalus*, crushing the juvenile shells. When the researchers removed some *Balanus* from an area in the lower intertidal zone, they observed that *Chthamalus* would colonize it without issue. Thus, it was the direct effect of *Balanus* on individuals of *Chthamalus*, namely squeezing out their juveniles trying to establish themselves, that gave *Balanus* the edge in this competition.

The barnacle study also illustrates one of the kinds of evidence for competitive exclusion that we see in nature: **ecological release**. If competitive exclusion is occurring, we would expect that removing one competitor would allow the other to move into the niche formerly occupied by the removed competitor. Alternatively, if two species exhibit competitive exclusion in one place, we would expect that a location where one species is absent would allow the other to expand its niche.

An example of this comes from lungless salamanders (Genus *Plethodon*). Lungless salamanders are small and have a long, thin shape that provides surface area so they can do all of their gas exchange through their skin, thus allowing them to live without lungs, as their name implies. One of their greatest centers of diversity is in the mountains of the Appalachians in the eastern United States.

Two species, *Plethodon jordani* and *Plethodon glutinosus*, live at different elevations on those mountains where they are both present, with *P. jordani* higher up and *P. glutinosus* lower down. *P. jordani* is not present on some mountains, and in those cases the range of *P. glutinosus* extends into the higher elevations where *P. jordani* is usually found. The absence of *P. jordani* releases *P. glutinosus* from the ecological competitive constraint on its range, allowing it to live farther up the mountain.

The other important piece of evidence for competitive exclusion is **character displacement**. Recall that in our hypothetical fruit fly community one species had to evolve in order to avoid competition. This was required because the two species were sympatric, but if the two species were allopatric there would be no need for the two species to differentiate their niches, even if the habitats were identical. It is often the case that two similar (and potentially competing) species do not have exactly the same geographic range, such that there are places where they are sympatric and others where they are allopatric. In sympatry, these species would have to differentiate their niches to coexist, but in allopatry they would not. Therefore, we would predict that if we compared individuals of these two species, they would be more similar when the individuals came from places where only one species was present, and they would be more different when they came from places where the species were sympatric.

We can illustrate this with an example of two species of snails in the genus *Hydrobia*. An important determinant of the snails' ecology is their body size, which is reflected in the size of their shells. The two species, *H. ulvae* and *H. ventrosa*, have overlapping ranges, so in some places the species are sympatric, and in other places only one species is present. Figure 6.8 shows the distribution of shell sizes for the two species when they are sympatric and when they are allopatric. Notice that one of the four graphs shows the snails to be quite a bit larger than those in the other three, and that these snails are *H. ulvae* when sympatric with *H. ventrosa*. In areas without *H. ventrosa*, the size of *H. ulvae* is similar to that of *H. ventrosa*. This suggests that populations of *H. ulvae* evolve to become larger when they are sympatric with *H. ventrosa*, and therefore avoid competition.

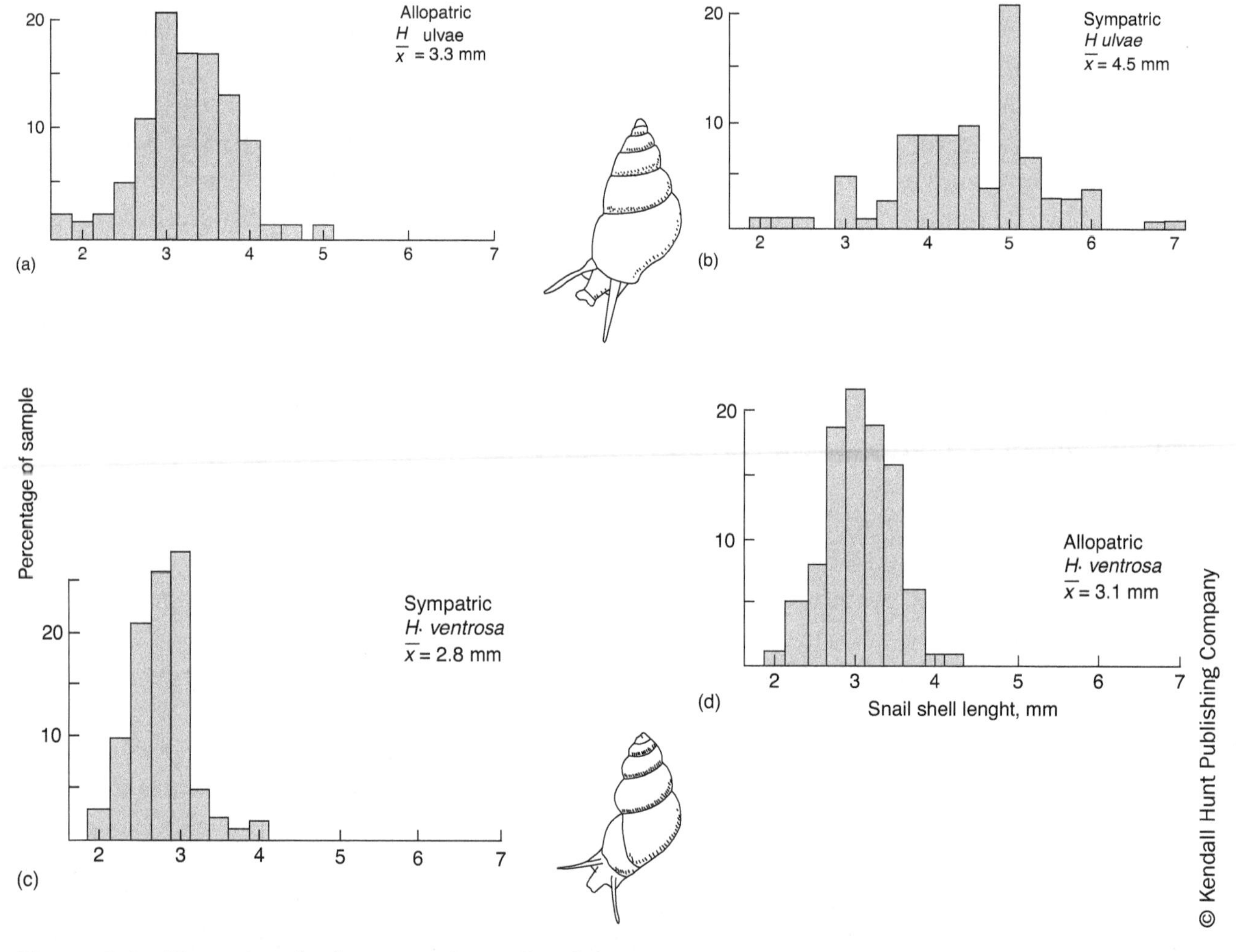

Figure 6.8. Character displacement is snails of the genus *Hydrobia*. *Hydrobia ulvae* is pictured on the left and *Hydrobia ventrosa* is on the right. The graph shows the distribution of shell sizes in each species when they are allopatric (A) or sympatric (S).

6.6 Biogeography

Geographic patterns were perhaps the most influential phenomena for the development of Darwin's ideas, and they are some of the strongest evidence for macroevolution. There is even an entire field of study about geographic patterns. **Biogeography** is the study of the patterns of geographic distribution of organisms and the processes that lead to it. An example of a question that a biogeographer might ask is: Why do polar bears (*Ursus maritimus*) live in the Arctic?

In fact, there are really two things about the current range of polar bears that need explaining. One is why they live in the cold region of the Arctic rather than warmer regions further south. To answer this question, we would likely explain it in terms of ecology and natural selection: polar bears are well adapted to cold climates but not to warmer ones, so selection favors their existence in the Arctic but not in warmer regions. These kinds of explanations come from the study of **ecological biogeography**.

But the Arctic is only one part of the world that is very cold, so this leads to the second fact that needs explaining, namely why polar bears live in the Arctic but not, say, Antarctica. Rather than an ecological explanation, this fact requires a historical explanation: polar bears don't live in Antarctica because they never got there. The ancestors of polar bears lived in the Northern Hemisphere, and polar bears arose as their cold-adapted descendants that could live in the polar regions of that hemisphere. Polar bears could not expand their range to Antarctica, because they would have had to expand their range into warmer regions to which they are not well adapted. Thus, the presence of polar bears in the Arctic and not in Antarctica can be explained as a phylogenetic "accident": the ancestors of polar bears happened to live in the Northern Hemisphere, but not in the Southern Hemisphere. These kinds of explanations come from the study of **historical biogeography**.

Biogeography has a close link to the history of evolutionary thought. Darwin was of course influenced by the distribution of animals on the Galápagos Islands, as well as on mainland South America. Alfred Russell Wallace, who essentially codiscovered Darwin's ideas, is often deemed the "father of biogeography" because he also was influenced by geographic distributions, particularly that of animals in the Indonesian archipelago. Evolution has thus long been considered important for understanding biogeographic patterns.

6.6.1 *Biogeography and trees: historical biogeography and dispersal*

There are two general processes that we can propose as historical biogeographic explanations. One is that organisms expand their range from an area where they were already present to another where they were not present. This process is called **dispersal**. The other explanation is that the geography itself changed, and the organisms that were present in that area became redistributed with those changes. This kind of process is called **vicariance**. Note that dispersal is explained by the organisms actively changing their distribution, whereas in vicariance the organisms are simply "going along for the ride" and do not have an active role in changing their distribution.

We have already seen an example of dispersal in the Hawaiian fruit flies: when new islands arose, the flies had to get to the new islands to expand their distribution. The change in geographic distribution also resulted in a specific pattern of speciation that was reflected in their phylogeny. We can use the same logical comparison of phylogenies and geographic distribution to determine patterns of dispersal through time.

Activity

Figure 6.9 shows a phylogeny of eight taxa that includes information of their distribution. The color of each terminal taxon (red or black) indicates where a taxon is found. The branches of the phylogeny are also color-coded to reflect how geographic distribution changed over the course of the group's evolutionary history. Let's assume that the geographic areas have not changed their position, although they may have changed their connection to each other at various times.

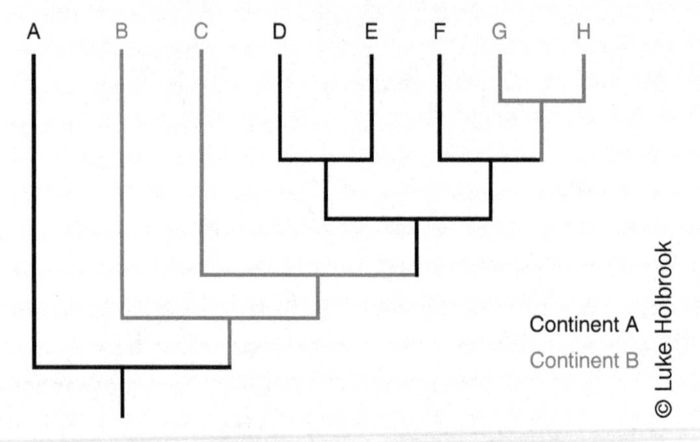

Figure 6.9. A phylogeny of eight taxa, which are colored red or black based on where they are found. The branches of the tree are also colored to indicate where a given lineage was evolving.

1. *Based on the phylogeny, how many times did these taxa disperse between the two continents during their evolution?*
2. *How many times does it appear that taxa dispersed from Continent A to Continent B?*
3. *How many times does it appear that taxa dispersed from Continent B to Continent A?*

Just like when we count the steps in character state changes, we count dispersal events as changes on the tree. In this case, there are three total dispersal events, two from black to red and one from red to black. We would therefore interpret this as the entire group ancestrally starting on continent A, then the ancestor of everything except taxon A dispersing to continent B, then the ancestor of taxa D, E, F, G, and H dispersing back to continent A, and finally the ancestor of taxa G and H dispersing to continent B.

6.6.2 Vicariance biogeography and continental drift

Vicariance often involves an area splitting into two or more fragments and taxa living in the original area becoming distributed among the area fragments, perhaps resulting in allopatric speciation. There are a number of processes that could cause an area to split up (and notice that by "area" we could be talking about an area of land or of water). We've already discussed some possibilities in our discussion of geographic isolation in allopatric speciation, and these are processes acting on a fairly local scale: a river cutting through an area, or a forest breaking up into smaller forest fragments due to climate change. But there are also geological processes that act on larger scales. **Plate tectonics** refers to the processes that drive changes in the configuration of the plates that make up the crust of the earth. The earth's crust can be divided into a number of independent plates, like tiles on a floor. Unlike floor tiles, plates are dynamic, where some plate margins are adding to the plate's edge, and other margins may be slipping below that of another plate to have their rock melted back into the mantle (Figure 6.10). This results in things that are on top of the plates, particularly continents, changing their positions over time.

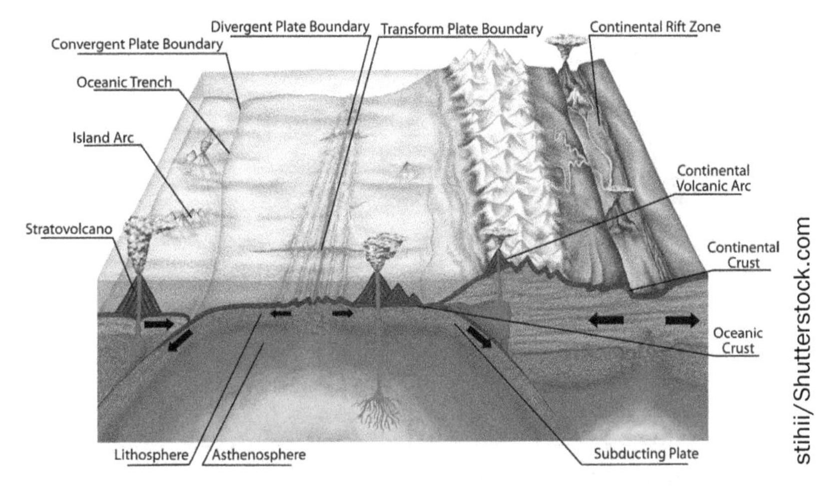

Figure 6.10. An illustration of the processes of plate tectonics. Note the arrows indicating when plates can move toward or away from each other.

The actions of plates explain a great deal about the most dramatic geological processes, like earthquakes and volcanoes, but they also explain the history of the face of the earth. The continents lie on top of plates, and their positions have changed as tectonic processes have moved plates relative to each other, a phenomenon we call **continental drift**. Continental drift had been proposed in the early twentieth century as an explanation for how various groups of organisms became distributed in disparate parts of the world. When first proposed, continental drift seemed absurd, and instead many palaeontologists proposed a variety of long-lost land bridges running between continents to explain similarities in the fauna and flora of two continents. It was only the discovery of plate tectonics through study of the sea floor that established that continental drift had indeed occurred in the past, and is still occuring. As a result, we now have unequivocal evidence that continents have divided and coalesced and divided again throughout the history of the earth, often changing shape in the process, with consequences for the organisms living in their terrain. Figure 6.11 illustrates the positions of the continents at various points in earth history.

We can test the influence of tectonic changes on patterns of organism distribution by comparing patterns of continental drift to the evolutionary history of organisms on those continents. When the discovery of plate tectonics in the 1960s established the validity of

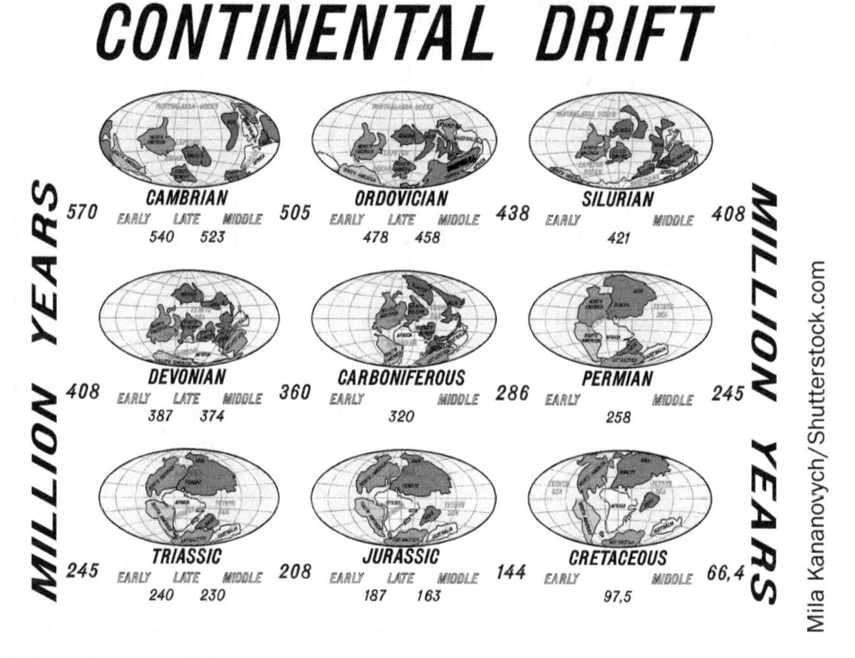

Figure 6.11. Changes in the positions of the continents at various times in earth history.

continental drift, palaeontologists realized there were many problems of historical biogeography that could now be solved. Some of the most interesting patterns in both living and extinct organisms concerned a collection of mainly southern land masses, including the continents of Australia, Africa, and South America, as well as the subcontinent of India and the large island of Madagascar. Today, we can see that Australia and South America are the centers of diversity for marsupials, "pouched" mammals like kangaroos, koalas, and opossums. Southern beech trees of the genus *Nothofagus* are also found in southern South America and in Australia and on its nearby islands. In the fossil record, there were a number of taxa known from two or more of these southern land masses, including a mammal-like "reptile" called *Lystrosaurus*, known from the Permian and Triassic (around 250 million years ago) of South Africa and India.

Reconstruction of the movements of the continents also suggested that these southern land masses were once connected with each other and another continent, Antarctica, in a supercontinent christened Gondwana. The notion that Antarctica once was positioned more north, in habitable climes compared to its current frigid location, and that it connected some of these continents, led palaeontologists to look for fossils in what is today the southernmost continent, and which is devoid of life beyond organisms that can survive on its coasts. Fossils from Antarctica include many of the taxa we would predict based on what is known from the rest of Gondwanaland: marsupials, *Nothofagus*, and *Lystrosaurus* have all been found there. Thus, marsupials were once spread across the combined expanse of South America, Antarctica, and Australia. (In fact, they were once present on all continents, with the possible exception of Africa, but went extinct on northern continents tens of millions of years ago.) Eventually, these three land masses broke apart and spread in different directions, and

Antarctica's movement to the South Pole doomed its terrestrial fauna to extinction, making the separation of marsupials on the other two continents more striking. (The opossum of North America is a relatively recent immigrant, crossing the Isthmus of Panama from South America about 600,000 years ago.)

As with dispersal, we can use phylogenies to test hypotheses of how vicariance might have affected the distribution and diversification of taxa. Vicariance involves an area splitting and taxa living in the original area becoming distributed among the area fragments, perhaps resulting in allopatric speciation. Figure 6.12 illustrates one way that we might expect phylogeny and historical geography to correspond when an area splits in this way. If an area goes through a process of splitting into smaller subareas, we would expect that the pattern of relationships among the organisms would match the pattern of splitting. Notice that in this example the organisms are not really going anywhere; where they end up depends on the geological processes causing the division. However, they are evolving, and geographic isolation results in speciation.

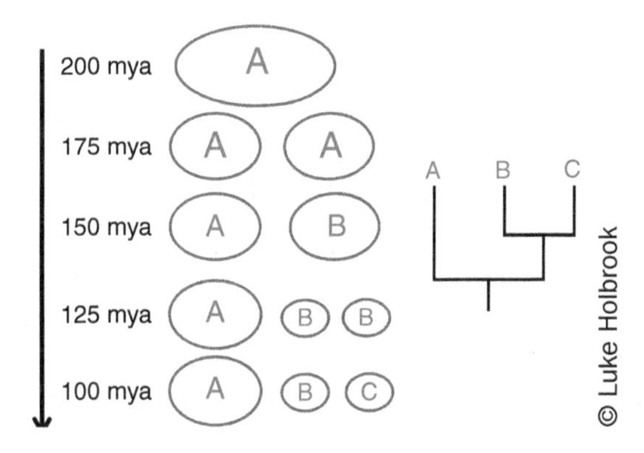

Figure 6.12. An illustration of vicariance and its effects on speciation. The letters represent species and the ovals represent areas where different species are found as the original area breaks up into smaller areas. The phylogeny of the three species is given on the right.

We can see this in the species on either side of the Isthmus of Panama. The Isthmus of Panama connected North and South America at least three million years ago. While this facilitated the movement of terrestrial organisms between the two continents, it created a barrier in the midst of the marine environment that had formerly been continuous between the continents, such that the Pacific Ocean was now separated from the Caribbean Sea. Thus, marine species that once ranged across the formerly continuous body of water were now separated into separate populations on either side of the isthmus. This resulted in allopatric speciation, and researchers have found that for many groups of organisms, the closest relative of a species on one side of the isthmus is a species on the other side.

6.6.3 Ecological biogeography: island biogeography

How and why specific taxa are found in a specific place is often a historical question, but ecological biogeography often focuses on more general patterns of diversity. For instance, why do islands have fewer species than continents, and why do big islands have more species than small islands? Note that what we mean by **diversity** is the number of taxa—usually the number of species—that we find in a location. So, two islands might have an equal number of individual mice populating them, but if the mice on one island all belong to the same species and the mice on the other island belong to 10 different species, the second island would have greater diversity of mice.

The biota of an island (i.e., the organisms that live on it) exhibits other interesting tendencies. Island organisms are more prone to becoming endangered or extinct: the poster child of extinction is an island bird, the dodo. Islands also often have high levels of **endemism**: in other words, they often have many species that are found on that island or archipelago and nowhere else (Figure 6.13). Why do islands exhibit these tendencies?

Figure 6.13. Two examples of the unsual life found on islands. (A) Lemurs, like these ring-tailed lemurs (*Lemur catta*), are found only on the island of Madagascar. (B) Giant tortoises are found in the wild today only on two archipelagos, the Galápagos Islands off the Pacific coast of Ecuador and the Aldabra Islands in the Indian Ocean.

What are the characteristics of islands that make them different from continents? Our questions already hint at an important factor: size. Islands are universally smaller than continents. There is another property of islands that is important for this discussion, particularly when we are considering oceanic islands, like the islands of Hawaii: they vary in **remoteness**, that is, how far they are from continents, or even from each other. So, how do these features contribute to the low diversity, high extinction, and high endemism of islands?

It is probably easiest to explain how remoteness contributes to these attributes of islands. An oceanic island that begins as a volcano emerging from the ocean is lifeless. Life on the island must first arrive as colonists from continents or other islands, and, obviously, the farther away an island is from sources of species, the more slowly it will accumulate them.

The size of the island might intuitively seem important for the number of species it can hold. Certainly, there is a well-documented relationship between the size of an island and the number of species on it. In fact, this relationship applies more generally, even when we look at different spatial scales on continents. The number of species of, say, birds you can find increases as you move from looking just at your backyard, to looking at your county, to looking at your state, to looking at the whole continent. But on islands the diversity is often low even compared to a comparable area on a continent. What is it about the small size of islands that reduces their diversity? It might be tempting to assume that small islands have fewer niches than big islands, and therefore fewer species can "fit" on a small island. But even islands of different sizes that have the same environments show this relationship.

The answer to this question of why islands have low diversity was finally produced in the 1960s by Robert MacArthur and Edward O. Wilson, who proposed the **equilibrium model of island biogeography**. Essentially, they said the diversity of an island is the result of two things, the number of colonizations and the number of extinctions; the difference between these two would determine the diversity on an island. Remoteness would have the biggest effect on colonizations, as it would be harder for new species to reach more distant islands. Extinction, on the other hand, was ultimately a product of population size: smaller populations would be at greater risk of extinction. The size of the population of a given species that an island can support is ultimately determined by its size. So, size determines diversity not by determining the number of niches, but by determining the population size of a given species, and therefore the risk of its extinction. MacArthur and Wilson tested their model by fumigating small mangrove islands of different sizes and distances from the Florida Keys, and then tracked the species that accumulated on the islands over time. As predicted, more remote islands tended not to be colonized as often, and smaller islands tended to have more species go extinct after colonization.

Literature Cited

Connell, J H. 1961. "The Influence of Interspecific Competition and Other Factors on the Distribution of the Barnacle *Chthamalus Stellatus.*" *Ecology* 42:710–23.

Roca, A L., N. Georgiadis, J. Pecon-Slattery, and S J. O'Brien. 2001. "Genetic Evidence for Two Species of Elephant in Africa." *Science* 293:1473-77.

Further Reading

Hurt, C., A. Anker, and N. Knowlton. 2009. "A Multilocus Test of Simultaneous Divergence Across the Isthmus of Panama Using Snapping Shrimp in the Genus *Alpheus.*" *Evolution* 63:514–30.

Mayr, E. 1991. *One Long Argument.* Harvard University Press, Cambridge, MA.

Part IV

Investigating the History of Life

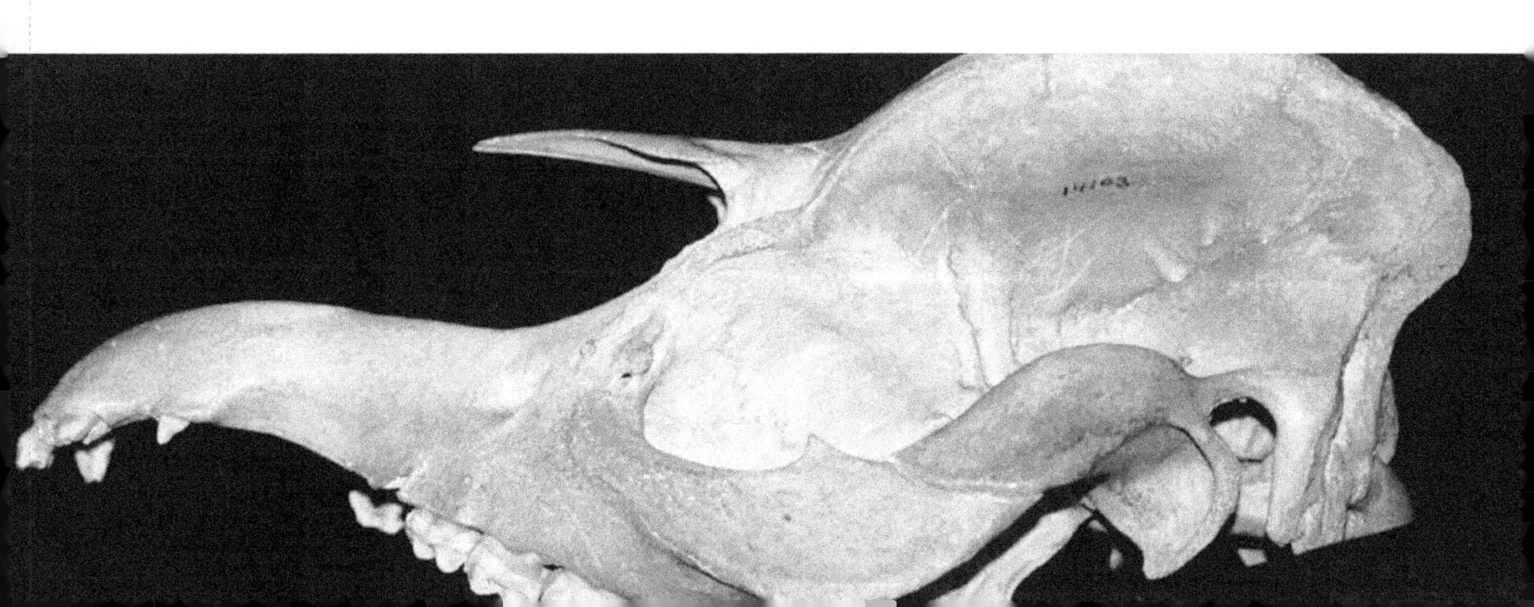

CHAPTER 7 Earth History and Patterns of Diversity

In Chapters 7–9, we will briefly survey the history and diversity of life, and Chapter 7 will provide some context regarding how the earth and the life on its surface have changed over time, and how we gather the evidence for this.

7.1 Determining ages of rocks and fossils

In Chapter 2, we discussed how we know the age of different rocks, and by extension the fossils they might contain, as well as how we know the age of the earth itself. Specifically, we discussed radiometric dating in detail, which is generally used on **igneous** rocks (Figure 7.1), rocks that are the immediate product of the cooling of magma. Another method that can be applied to igneous rocks is **paleomagnetism**, which looks at the orientation of iron crystals in such rocks. Iron crystals will align themselves according to the orientation of the earth's magnetic field while the magma is still fluid. The magnetic field has actually changed frequently in earth history, such that the magnetic North Pole has frequently "flipped" to be at the geographic South Pole. The paleomagnetic record does not produce absolute dates in and of itself, but we can use the pattern of magnetic "normal" and "reversals" to align rocks from different areas according to the magnetic signatures they bear.

While we can determine the ages of igneous rocks with these methods, igneous rocks do not preserve fossils, as the remains of living organisms are destroyed by the heat of the magma. Fossils are found almost exclusively in **sedimentary** rocks, which are typically formed from sediments that are deposited in aquatic environments and that are later cemented together and hardened to form rock. Sediments come from the erosion of other rocks; so, while we could measure radioactive isotopes in a sedimentary rock, it would not tell us the age of the sedimentary rock but instead the age of the source rocks that were eroded. Given that we are interested in the ages of fossils, how do we determine the ages of sedimentary rocks that preserve them?

In Chapter 2, we mentioned briefly that, besides methods of absolute dating like radiometric dating, there are also ways to determine the relative ages of rocks, based on their relations to one another. The most basic principle for accomplishing this is the **law of superposition**, which essentially states that we expect younger rocks to be deposited on top of older rocks. In places where geological processes have not deformed the landscape and where water has exposed rock layers below the surface, such as the Grand Canyon and many other parts of the western United States (Figure 7.2), it is relatively easy to apply this principle to visible rock layers, or **strata**. In other places, strata have been altered and reshaped by folding, faults, erosion, and other processes, and determining which layers are younger and which

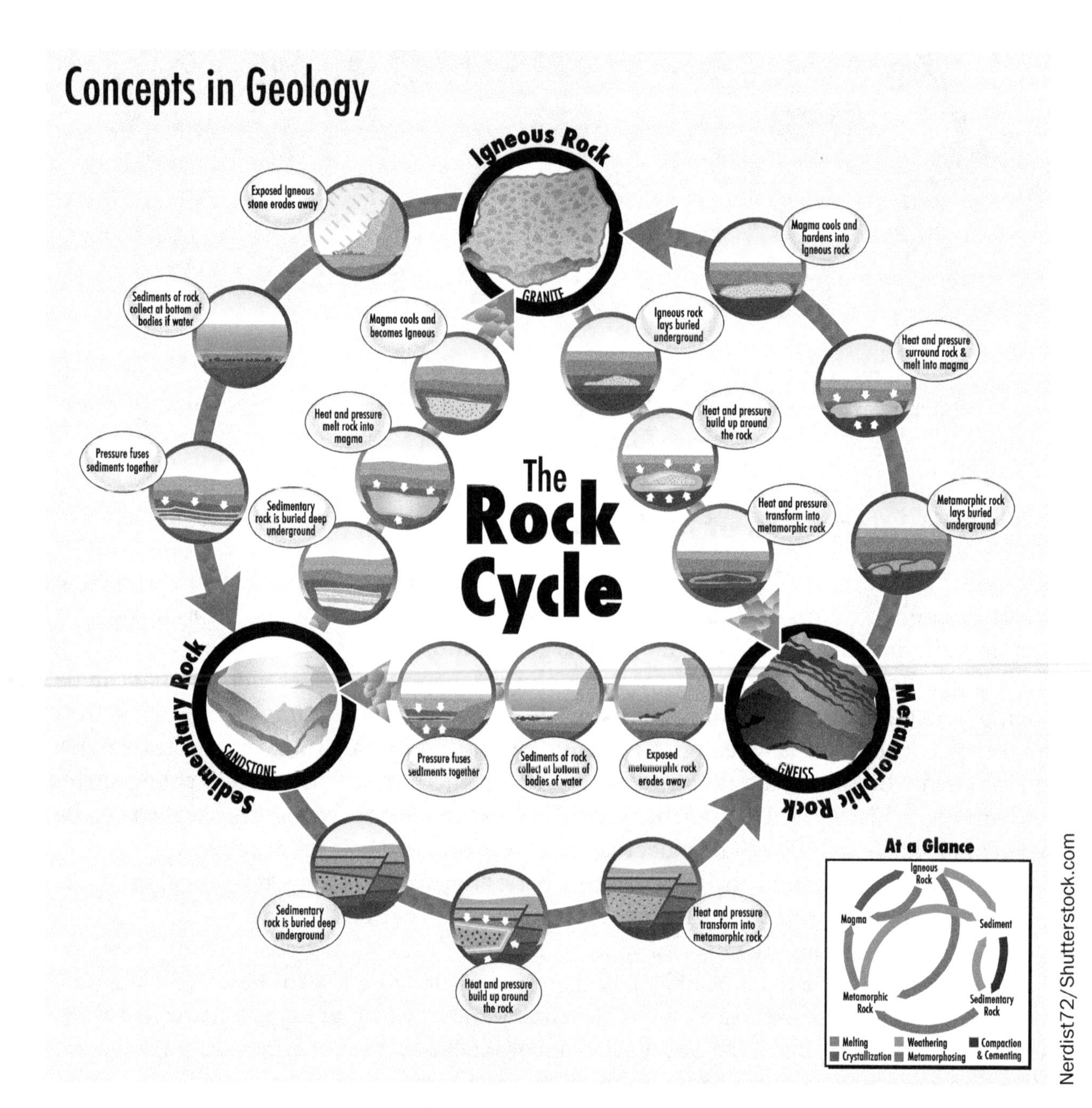

Figure 7.1. Types of rocks and the rock cycle.

are older requires puzzling out this history. This is the work of the geological discipline of **stratigraphy**. While stratigraphy itself tells us relative ages of rocks, if these rocks include igneous rocks with dates, then we can constrain the ages of other rocks, such as sedimentary rocks, based on their relationships to the dated igneous rocks.

In many cases, the fossils preserved in sedimentary rocks can also help to determine their age. For instance, if we can establish the age of a certain fossil in a certain location based on the stratigraphy of the sedimentary rock in which it is preserved, then if we find that fossil in other rocks whose ages are not constrained by igneous rocks, we might be able to infer that rocks preserving this fossil are the same age. Stratigraphy that incorporates fossils in this way is called **biostratigraphy**.

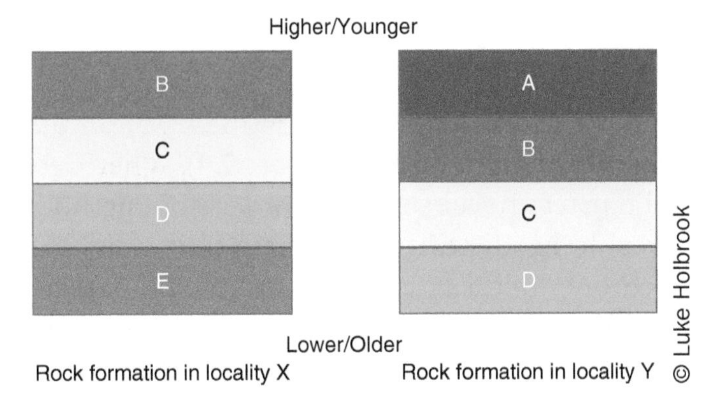

Figure 7.2. An illustration of the principle of superposition. Each colored block represents a layer of rock found in that locality. Older layers are lower in the rock formation and younger layers are higher. The colors and letters represent layers with similar fossils and that therefore are interpreted as being of similar ages.

7.2 Fossils

A **fossil** is a remnant of an organism preserved in the geological record. Most fossils are **body fossils** (Figure 7.3), actual parts of an organism that have been preserved. Typically, these are parts of an organism that are hard, durable, and mineralized, such as bones, teeth, and shells. The minerals in a body fossil, such as the calcium carbonate in a shell, are often replaced over time by the minerals in the rocks that hold them, changing their color and weight.

Fossils can also come in the form of **trace fossils** and **chemical fossils** (Figure 7.3). Trace fossils are produced by organisms but are not actual parts of them. These would include burrows, footprints, and trackways. Chemical fossils are molecules of biological origin preserved in rocks. Chemical fossils have been of particular interest for studying the earliest evolution of life, during which life consisted exclusively of microbes that generally do not leave body fossils but that might leave chemical signatures of life in the rocks.

Figure 7.3. Fossils. (A) A body fossil of a fish. (B) A trace fossil of a dinosaur footprint.

7.3 The geological timescale

Once the ages of rocks and their fossils can be determined, we can use this information to develop a **geological timescale** of earth history (Figure 7.4). When people first started to do this in the eighteenth and nineteenth centuries, there were no methods for absolute dating, but rocks of similar ages could be identified, and these time units based on similarly aged rocks could be placed in chronological order based on relative dating. This is where we get the various named time periods that you can see in the geological timescale in Figure 7.4. Some of the names come from places where the rocks of that age were first recognized (e.g., the Devonian is named for Devon in England) or from other characteristics of the rocks (e.g., the Cretaceous is named for the Greek word for chalk, *kreta*, in referenced to chalk deposits from that time, like the White Cliffs of Dover in England).

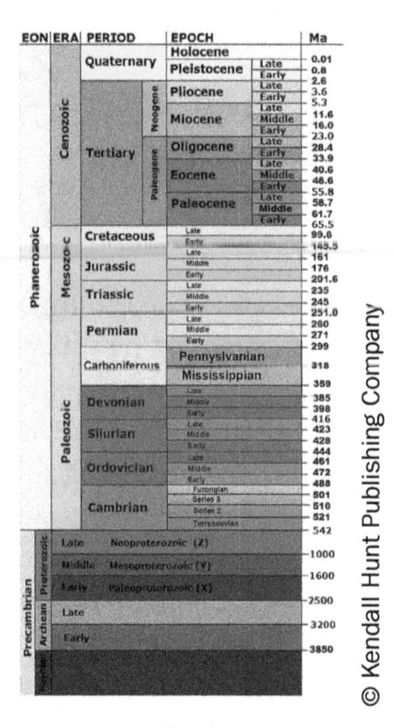

Figure 7.4. The geologic timescale.

Each named time period covers some expanse of time, and subsequent study has often allowed us to subdivide these time periods. As a result, the various named time periods form a nested hierarchy, just like in the classification of organisms. The largest time units are **eons**, and each eon is subdivided into several **eras**, which are further subdivided into **periods**, which themselves are divided into **epochs**.

You may have noticed from Figure 7.4 that named time units of the same rank do not necessarily have the same duration. The **Cenozoic** era, for instance, covers only the last 65 million years, whereas the **Mesozoic** and **Paleozoic** eras lasted 180 million and nearly 300 million years, respectively. The reason for this kind of disparity has to do with how these rock units were identified historically, and this relates to how well we know different time periods from the rocks, and how much of the geologic record preserves rocks of those ages. The older a rock is, the more time there is for it to be eroded or destroyed by other processes, and the less likely it is that fossils in those rocks will survive to the present. As a result, we know a great deal more about the Cenozoic and its fossils than we do of the Paleozoic and its fossils. Furthermore, organisms that can produce fossils, that are more likely to produce fossils, or that have fossils that are more easily recognized have not been present throughout earth history. The Paleozoic, Mesozoic, and Cenozoic are the eras of the **Phanerozoic eon**, which covers the last 540 or so million years, and which is the eon in which the vast majority of fossils are found. Besides the fact that the Phanerozoic is more recent than other eons, it also appears to mark the beginning of abundant life with hard parts that can be fossilized.

7.4 Origination, diversification, and extinction

Once we have a timescale for earth history and the history of life, we can examine patterns in how diversity of life or of specific taxa changes over time. Paleontologists are often interested in the **origination** of a taxon, or when a lineage first appeared. The earliest fossils of a group give us some information on its time of origination, but, because of the incompleteness of the fossil record, the earliest fossils of a group are almost certainly younger than the group's actual origin. This is often even more apparent when we consider that a lineage becomes distinct when it diverges from the ancestor it shared with its closest relative. That divergence has to be older than the oldest fossils both of the group in question and of its oldest relative. From this information, when a taxon's earliest fossils are much younger than those of its closest relative, we can infer a gap in the fossil record, called a **ghost lineage**.

Many groups exhibit **diversification**, or an increase in diversity, after their origin; in some cases, we can hypothesize reasons for this. In more recent cases, we can infer that a taxon that finds itself in an environment with little competition from other taxa might undergo an **adaptive radiation** (Figure 7.5), where an ancestral species diversifies into multiple new species that adapt to empty niches. We have encountered an example of this in earlier chapters with the Galápagos finches and Hawaiian fruit flies, and there are other adaptive radiations related to colonizing islands, such as Hawaiian honeycreepers, as well as lemurs and other vertebrates on Madagascar. The diversification of placental mammals after the extinction of the dinosaurs is another case thought to represent an adaptive radiation.

The adaptive radiation of placental mammals was facilitated by an **extinction**, essentially when a taxon's diversity becomes zero. Extinction is a fairly common phenomenon in earth history. Most extinctions occur randomly through time, often of single species at a time, and these extinctions comprise what is known as **background extinction**. Occasionally, there are times when many taxa go extinct at the same time, and these are known as **mass**

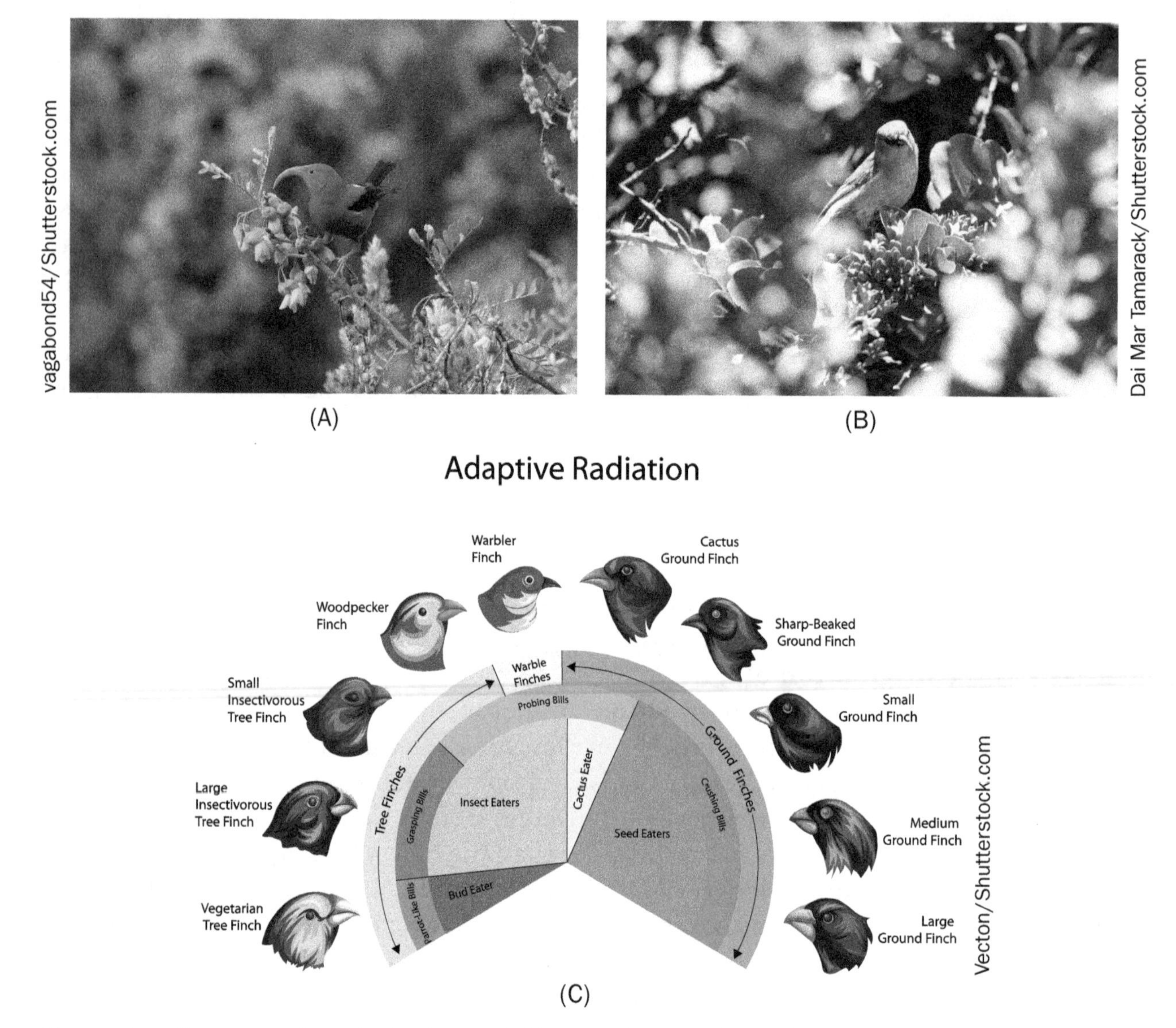

Figure 7.5. Two examples of adaptive radiations on islands. (A) and (B) Two examples of Hawaiian honeycreepers. (C) The radiation of Galápagos finches.

extinctions. Paleontologists recognize five major mass extinctions, the most famous of which is the extinction at the end of the Cretaceous or at the boundary between the Mesozoic and Cenozoic. Because the abbreviation for the Cretaceous is K and the first part of the Cenozoic is called the Tertiary (T) or the Paleogene (Pg), this extinction is often referred to as the **K–T** or **K–Pg extinction**. Many groups of organisms experienced extinction at this time, but this extinction is best known for wiping out the dinosaurs (other than birds). Yet, the K-Pg extinction is not the biggest mass extinction. That title goes to the extinction at the end of the Permian, at the boundary between the Paleozoic and the Mesozoic, and it is estimated that as much as 95% of species alive at that time went extinct.

The causes of mass extinctions have to be something with a global effect to explain a global pattern. These can include climate change and its effects, such as changes in sea level, as well as environmental changes associated with tectonic events, such as the impact of the formation of Pangaea, where all of the continental land masses were essentially combined into one supercontinent. Such events likely occurred over relatively long periods of time, even if they appear as instants in geological time.

Other potential causes would have been shorter in duration and affected the life relatively quickly. The best-known example of this is the impact of a **bolide**—an enormous extraterrestrial object, such as an asteroid—off the coast of the Yucatan Peninsula that is thought to have been the cause of the K-Pg extinction due to rapid, short-term, catastrophic environmental changes that resulted from the impact. The evidence for this impact and its global effect first came to light with the discovery of unusually large amounts of iridium—an element that is rare in the earth's crust but more common in extraterrestrial objects—in rock layers that mark the K-Pg boundary. Later investigations of the K-Pg boundary added to this evidence with the discovery of **shocked quartz**, a characteristic type of quartz that occurs at known impact sites, and **glass spherules** that result from the melting of silica-rich rocks due to the heat from the impact. The fact of the bolide impact at the K-Pg boundary is well established, including the identification of the impact site, but while we know that the impact happened at the time of the extinction, there is still work to be done to establish the causal link of the impact and the extinction and to determine how exactly the impact caused the extinction of so many taxa on the land and in the seas.

We are currently in the midst of what has been called the sixth major mass extinction in earth history, which has resulted in the loss of species at a rate unprecedented in the history of mass extinctions. These extinctions are clearly linked to human agency, either due to overhunting, habitat loss, introduced species, climate change, or other human impacts on the environment. This is not the first time that humans have had a hand in extinctions. Many mammals and birds, particularly large species—also called **megafauna**—went extinct at the end of the Pleistocene. This Pleistocene is well known for the series of **glaciations**, or "ice ages," that occurred during it, with each glaciation bounded by **interglacial** periods (including which we are living in now). Thus, the Pleistocene was a time of great climatic fluctuations, but the extinctions did not occur with each glaciation. Rather, extinction occurred on different continents and islands at different times between about 50,000 and 10,000 years ago, and in many cases the extinctions coincide with the arrival of humans. The case for humans as the sole or main cause of extinctions varies by location: the evidence for human agency is strong for Australia but weak for Eurasia, and it is likely that the extinctions in different locations were caused by varying mixtures of human impact (presumably hunting, or **overkill**) and climate change. The late Pleistocene extinctions are not themselves counted among the five major extinctions, but they are so close to today in geologic time that they might as well be lumped into the current extinction crisis.

7.5 The first four billion years

The Phanerozoic covers a little more than the last half billion years of earth history, which means that more than four billion years of earth's 4.6 billion year history preceded this eon. The first four billion years of earth history is divided into three more eons, each longer than the Phanerozoic. The shortest of these is the Hadean, which covers the first 600 million years and essentially the only part of earth history without life. The next 1.5 billion years comprise the Archean, which includes the earliest evidence of life. Finally, the Proterozoic extends for about two billion years from about 2.5 billion years ago until the beginning of the Phanerozoic, during which we see the earliest eukaryotes and the first multicellular life.

In its early, lifeless ages, the earth's surface was highly energetic. The surface was volcanically active, the early atmosphere was stormy, and there was no ozone layer to filter out ultraviolet (UV) radiation. Liquid water accumulated to form the first oceans around 4.4 billion years ago. The earliest fossil evidence of life goes back to 3.8 billion years ago, in the form of chemical fossils interpreted as products of microbes, and the earliest unequivocal trace fossils of life are stromatolites, mounds of layered sediment produced by mats of microbes, that are 3.5 billion years old (Figure 7.6).

Figure 7.6. Stromatolites. (A) A fossil stromatolite. (B) Stromatolites found today in waters near Australia.

The microbes that formed those stromatolites were likely **photosynthetic**; that is, they fueled their cellular metabolic processes by the conversion of carbon dioxide and water into simple sugars, using energy from sunlight. A by-product of this is oxygen gas, which was essentially absent from the atmosphere of the earliest earth and would only start to accumulate after photosynthesis evolved. By 2.3 billion years ago, oxygen levels reached a point, though still far below the oxygen levels of today, that was high enough to leave its mark on the earth's atmosphere and geology, reducing the relative amount of greenhouse gases like methane to cool the earth, and oxidizing the iron-rich rocks on the surface. Geologists call this turning point the **Great Oxidation Event**.

Still represented only by single-celled organisms, life evolved to cope in a more oxygenated world, and it is thought that coping with the reactive nature of oxygen might have been an important driver in the evolution of eukaryotic organisms, as we shall discuss in a later chapter. The first possible evidence of multicellular organisms comes from fossils that appear to be small sponges that are more than 700 million years old, which also happens to be during a time of unusually cold global climate called the **Cryogenian**. After the Cryogenian ends around 635 million years ago, the first unequivocal evidence of multicellular life shows up in the form of the **Ediacaran** fauna from Australia, which is about 575 million years old (Figure 7.7). These organisms are known from impressions of their soft bodies, as none of these taxa have hard parts. In fact, it is difficult to say exactly what they are, as they show some similarities to animals, but some researchers consider them to be their own kingdom of multicellular organisms. The end of the Ediacaran Period marks the beginning of the Cambrian Period and the Phanerozoic Eon.

scigelova/Shutterstock.com

Figure 7.7. An example of an Ediacaran fossil.

7.6 Paleozoic seas and the first life on land

As mentioned before, most of what comprises the fossil record is known from the Phanerozoic, despite it being such a small part of earth history, because its rocks have survived, and because organisms with hard parts are found almost exclusively during this eon. By the middle of the Cambrian, the first period of the Paleozoic Era, we know of abundant communities of animals, including early representatives of sponges, corals, arthropods, echinoderms, mollusks, and even our own phylum of animals, the chordates. Like many of their descendants, all of these creatures were aquatic and usually marine. Some unusual fossil localities, such as the Burgess Shale in Canada (Figure 7.8), preserve soft-bodied animals and reveal an even greater diversity of animals, some of which are difficult to relate to phyla either alive today or known from hard parts in the Cambrian. The Cambrian animal diversity appears relatively suddenly in the fossil record, at least from a geological perspective, and the sudden appearance of these taxa is known as the **Cambrian Explosion**. There is some question as to what this "explosion" really was, and why it occurred. Paleontologists have interpreted this sudden appearance as a rapid diversification of animal life, perhaps in response to ecological changes in the ancient seas. Some molecular clock studies, on the other hand, suggest that animal phyla had begun to diversify well before the Cambrian, suggesting that

(A) (B)

Figure 7.8. (A) The locality of the Burgess Shale. (B) A reconstruction of some of the Cambrian life preserved in the Burgess Shale, including *Anomalocaris*.

the "explosion" was an artifact of the fossil record, but more recent studies have tended to converge on the "explosion" being a real diversification event.

The seas of the Paleozoic were populated with an abundance of animal life, including a variety of invertebrates and vertebrates. By the Devonian, diversity of fishes (and the absence to that point of terrestrial vertebrates) was great enough that this period is sometimes called "the Age of Fishes" (Figure 7.9). However, fossil invertebrates are even more diverse throughout the Paleozoic. Probably the most characteristic animals from this era are the **trilobites** (Figure 7.9), which first appear in the Cambrian and are very diverse into the Devonian, during which they decline, hanging on only to wink out of existence at the end of the Permian.

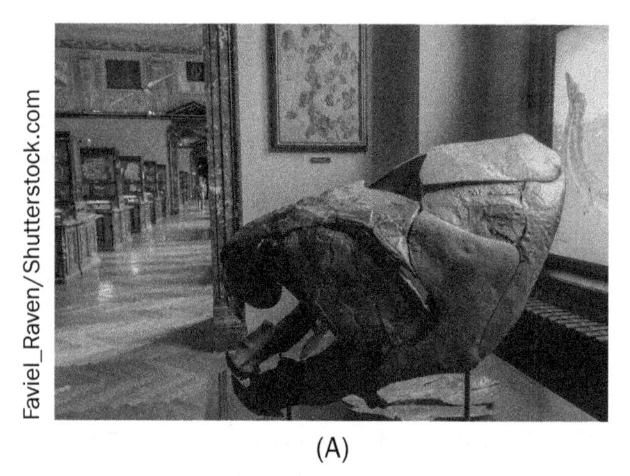

(A) (B)

Figure 7.9. (A) The skull of a giant predatory fossil fish from the Devonian, *Dunkleosteus*. (B) A trilobite.

7.6.1 The first life on land

The colonization of land actually begins as early as the Ordovician, when we find the first fossil evidence of plants. The first land animals appear in the late Silurian in the form of the earliest millipedes. Arthropods, especially arachnids and insects, have exploited the terrestrial world with great success and have done so for a longer time than any other animal phylum. The first vertebrates don't appear on land until the late Devonian, in the form of the earliest **tetrapods**, although the earliest evidence appears to be trackways from Poland that predate the earliest body fossils of tetrapods.

After the Devonian, life on land is quite rich, and we see evidence of the first forests in the Carboniferous. The Carboniferous gets its name from the coal deposits that are common from that time period, and which are derived from the decomposition of the rich plant life from that period. The Carboniferous is also when we see the first amniotes (relatives of "reptiles," birds, and mammals), which have diversified into distinct lineages related to today's amniote groups by the end of the Paleozoic. Perhaps due to the rich plant life, oxygen levels in the late Paleozoic reached unusually high levels, and these high oxygen levels are thought to have allowed the evolution of unusually large arthropods, which are constrained in their body size by the rate of diffusion of oxygen. Higher oxygen concentrations could support larger arthropods, such as the giant dragonfly *Meganeura*, which had a wingspan over 70 cm—more than two feet!—as well as large amphibians, many of which use their skin for gas exchange (Figure 7.10).

Marbury/Shutterstock.com

Figure 7.10. Reconstruction of the late Paleozoic giant fossil dragonfly *Meganeura* on the trunk of a tree.

7.6.2 The Permian extinction

The Paleozoic ends with the largest mass extinction ever, the **end-Permian** or **Permo-Triassic extinction**. As much as 95% of species alive at that time went extinct. The causes are still a bit murky, but the extinction coincides with some significant geological events. The end of the Paleozoic is marked by the formation of Pangaea, the supercontinent that collected all continental masses on the surface of the earth (Figure 7.11). The formation of Pangaea likely took too long to explain the relatively rapid mass extinction, but it would have had an impact on the available nearshore environments to support marine life, precipitation reaching the interior of the supercontinent, and possibly even the amount of oxygen in waters cut off from global circulation. The cause that is currently favored as most important

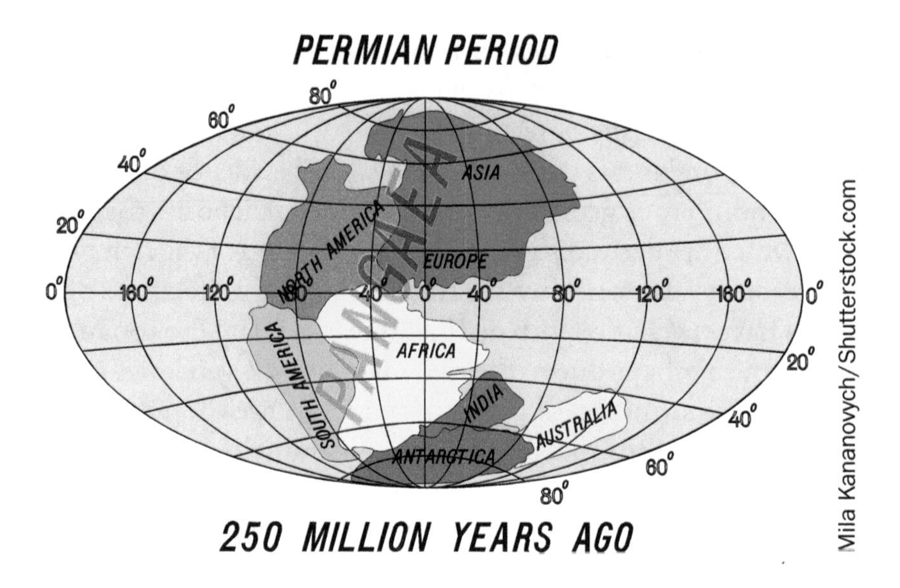

Figure 7.11. The supercontinent Pangaea as it was at the end of the Permian.

is the large amount of volcanism that occurred at this time, marked by the massive igneous rock formation in Siberia called the Siberian Traps. These extensive volcanoes would have introduced large amounts of gases that would both increase the greenhouse effect and contribute to acid rain, resulting in global warming, sea level rise, ocean acidification, and depletion of atmospheric oxygen. This combination of effects would challenge numerous marine species and reduce availability of suitable habitats for all life.

7.7 The age of dinosaurs

The Mesozoic is known as "the Age of Reptiles," primarily because it is those amniotes, especially dinosaurs, that dominated the terrestrial vertebrate faunas, and not mammals, which would dominate the Cenozoic. From a broader perspective, the Mesozoic really was when the modern fauna and flora originated, although we will say more about the modern biota in the discussion of the Cenozoic. The Mesozoic is when we see the first mammals, the first birds, the first turtles, the first snakes, and the first flowering plants. The seas of the Mesozoic also looked more like the seas of today, in that they were dominated by mollusks, crustaceans, and ray-finned fishes, rather than by trilobites, sea lilies, and brachiopods. All of these groups were present in the Paleozoic, but in the Mesozoic the groups of the modern fauna diversified greatly.

7.7.1 Dinosaurs

Dinosaurs (Figure 7.12) first appear in the late Triassic. Dinosaurs are **archosaurs**, a group that also includes crocodilians (crocodiles, alligators, and their kin) and **pterosaurs**, the first group of vertebrates to achieve powered flight. Birds are also archosaurs, but we now know that birds are dinosaurs; in fact, they are part of the group of dinosaurs called **theropods**,

Gilda Villarreal/Shutterstock.com

(A)

Dotted Yeti/Shutterstock.com

(B)

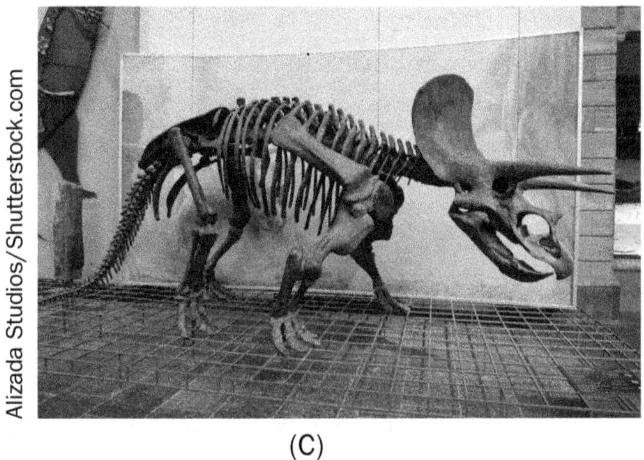

Alizada Studios/Shutterstock.com

(C)

Figure 7.12. Dinosaurs. (A) The skeleton of *Tyrannosaurus rex*. (B) A reconstruction of *Brachiosaurus altithorax*. (C) A skeleton of *Triceratops horridus*.

which includes familiar carnivorous dinosaurs like *Tyrannosaurus rex*, *Allosaurus*, and *Velociraptor*. Theropods are usually included in the **Saurischia** (literally "lizard-hipped") with the giant, herbivorous, long-necked dinosaurs called **sauropods**, which include what were the largest land animals ever known. Other dinosaurs are placed in the **Ornithischia** ("bird-hipped"), and these include ankylosaurs, ceratopsians like *Triceratops*, stegosaurs, and hadrosaurs or so-called "duck-billed dinosaurs," among others. Like sauropods, ornithischians are generally thought to be herbivores. There is some debate regarding the relationships among dinosaurs, and a recent study removed theropods from Saurischia and placed them with ornithischians, but many paleontologists have been skeptical about adopting this scheme.

Dinosaurs thrived for around 180 million years and dominated the Mesozoic landscape. They formed communities of herbivores and carnivores that are reminiscent of the mammalian communities we have today, although they were generally larger in body size. Large theropods and the larger sauropods and ornithischians were far larger than the largest modern mammalian carnivores and herbivores, respectively, and, excluding birds, the smallest dinosaurs known were still about the size of a chicken. The only dinosaurs that approach the small size of mammals like rodents and bats are modern songbirds and hummingbirds.

Research on dinosaurs over the past few decades belies the sluggish, unsophisticated reconstructions of earlier in the twentieth century. Dinosaurs appear to have been quite active creatures, and there is a lively debate regarding whether or not they were **endothermic**—that is, could they, like mammals, produce enough heat metabolically to maintain a high body temperature that would allow them to be active longer and in a broader range of climates? Of course, birds are endotherms, so at least some dinosaurs are endothermic, and endothermy at least must have evolved in the lineage leading to birds. But how early did endothermy evolve in dinosaur phylogeny? One of the most interesting developments in recent years is the discovery of evidence for feathers in many dinosaurs that are not birds. Feathers have functions in a number of contexts, including providing wing surfaces for flight, displaying colors and patterns to other individuals, and insulation for retaining the body heat produced by an endotherm. While feathers alone do not necessitate that a feathered dinosaur was an endotherm, they are highly suggestive of endothermy in those taxa that have them.

Other research has revealed that dinosaurs exhibited a variety of sophisticated behaviors more reminiscent of their bird descendants than of some of their more distant reptilian relations. Some dinosaurs show evidence of herding (or perhaps we should say flocking) or of parental care. Dinosaurs also lived in unexpected places, such as north of the Arctic Circle, at a time when temperatures were warmer, but still cooler than in the tropics.

7.7.2 Mesozoic seas

As mentioned before, in a broad sense, the seas of the Mesozoic were much like the seas of today, dominated by mollusks, crustaceans, and vertebrates, particularly ray-finned fishes. Despite this, Mesozoic sea life was still distinctly different from today. Much of North America was covered by a shallow inland sea, producing the Cretaceous chalks of modern Kansas. The mollusks of the Mesozoic included ammonoids and nautiloids, two groups of cephalopods that, unlike their squid and octopus relatives, had large, often coiled shells (Figure 7.13). Besides sharks and large ray-finned fishes, the large predators of the oceans included several kinds of marine reptiles, including plesiosaurs with sea lion-like flippers and some with long necks; ichthyosaurs with dolphin-like bodies and fish-like tail fins; and mosasaurs, giant lizards related to Komodo dragons with flippers and tail fins (Figure 7.13). In addition to these impressive marine reptiles, enormous crocodiles and giant sea turtles were also found in Mesozoic seas.

7.7.3 The K–Pg extinction

As we have mentioned already, the Mesozoic ends with a mass extinction at the end of the Cretaceous, the K-Pg extinction. About 65% of species disappeared, including all dinosaurs apart from birds; all of the ammonoids and almost all of the nautiloids, with only a handful of species of nautilus living today; and the last of the pterosaurs, plesiosaurs, and mosasaurs. Many other groups were severely depleted though not extinguished.

The cause of this extinction has been attributed to a variety of things, including climate change (specifically global cooling) and volcanism, but, as we discussed earlier, the evidence increasingly points to the impact of an extraterrestrial object as the trigger.

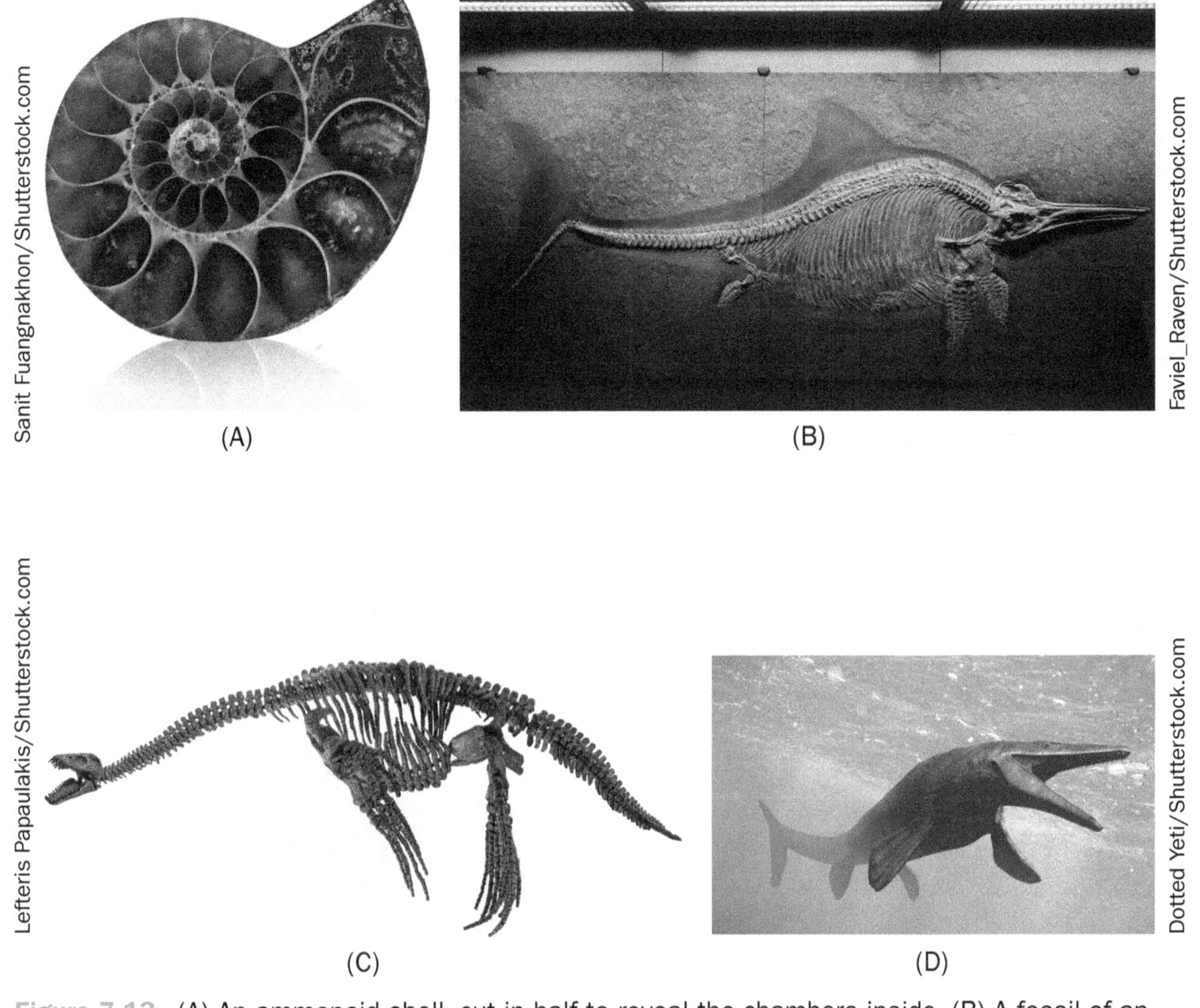

Figure 7.13. (A) An ammonoid shell, cut in half to reveal the chambers inside. (B) A fossil of an ichthyosaur, a Mesozoic marine reptile. (C) A fossil of a plesiosaur, a Mesozoic marine reptile. (D) Reconstruction of a mosasaur, a Mesozoic marine reptile.

7.8 The rise of the modern biota

In a broad sense, the major lineages of plants and animals that comprise our modern flora and fauna were in place by the end of the Cretaceous, but throughout the last 65 million years that biota has been reshaped to become what we recognize today. The most familiar change concerns the expansion of mammal diversity to occupy most of the ecological space opened up by the extinction of dinosaurs and other large reptiles. Mammals in the Mesozoic were represented largely by small, insectivorous to omnivorous forms, though recent discoveries have expanded the ecological diversity of these early mammals and their relatives: they include forms as large as a beaver, specialized diggers, semiaquatic forms, and even gliders.

7.8.1 From hothouse to cool and dry

Much of the shaping of the biological world during the Cenozoic can be attributed to two things: climate change and tectonics. Mesozoic global temperature peaked in the late Cretaceous and decreased into the early Cenozoic, yet temperatures were still much higher than they are today. In fact, temperatures climbed in the Paleocene epoch, and the boundary between the Paleocene and Eocene is marked by a **hyperthermal**, a spike in global temperature, known as the **Paleocene–Eocene Thermal Maximum** (**PETM**). Temperatures continued to climb in the early Eocene, then began a cooling trend that largely characterizes the rest of the Cenozoic. A sharp drop in temperature at the end of the Eocene marked the point when temperatures were cool enough for an ice cap to form on Antarctica, though it would thaw later and reform in the Miocene. The temperature gradient across latitudes became sharper than it was in the Mesozoic, resulting in more seasonal climates away from the equator.

The cooling during the Cenozoic allowed for the expansion of plants that would define a new kind of ecosystem: grasslands. Cooler and drier climates favored grasses over leafy plants, and therefore favored open habitats like prairies over closed habitats like forests. Grasslands led to the evolution of specialists in these habitats, such as large grazing herbivores with teeth specialized for eating tough grasses and limbs adapted for running, both for evading predators and for enduring long migrations (Figure 7.14).

Lifestyle_Studio/Shutterstock.com

Figure 7.14. Grasses.

7.8.2 Fragmenting continents

During the Mesozoic, Pangaea breaks into two supercontinents: **Laurasia**, consisting of North America and most of what today is Eurasia; and **Gondwana**, consisting of South America, Antarctica, Australia, Africa, Madagascar, and India. Throughout the Cenozoic, Laurasia stayed more or less intact, with North America, Europe, and Asia (minus India) being connected to or disconnected from one another at different times, largely due to changes in sea level.

By the beginning of the Cenozoic, Gondwana had begun to break up and its constituent parts migrated elsewhere. Antarctica eventually moved over the South Pole, effectively ending terrestrial life there. Africa, Madagascar, and India moved northwards, with Africa connecting to Eurasia via the Arabian Peninsula, and India "crashing" into Asia, with the Himalayas forming in the crumple zone of the impact. South America moved northwards but spent most of the Cenozoic as an island continent, isolated from the others until the Isthmus of Panama formed by three million years ago to connect it to North America.

These tectonic events had an enormous effect on redistributing life around the globe. The connections among the continents of Laurasia allowed taxa to migrate between them, which is why they share a lot of similarities in their fauna and floras. Fragments of Gondwana connected to parts of Laurasia and exchanged plants and animals between the areas. This allowed horses and rhinos to get to Africa and possibly for their ancestors to migrate from India to Asia.

Perhaps the most interesting example of faunal exchange between continents is between North and South America. Prior to connecting to North America, South America had a unique fauna: marsupials, including large carnivores and even a sabertooth marsupial; unusual large hoofed herbivores unique to that continent; anteaters, sloths (including giant ground sloths), armadillos, and their relatives; large guinea pig-like rodents, like porcupines and capybaras; and monkeys. A few of these taxa moved into North America, including one species of armadillo, one marsupial (the opossum), some giant ground sloths, some giant, tank-like armadillos called glyptodonts, and one species of porcupine. Quite a few species moved into South America from North America, including many that we associate with that continent today: jaguars and other cats (including a sabertooth species), tapirs, deer, foxes and wild dogs, peccaries, camels (including llamas), bats, and mice. There was even an elephant and a horse that migrated south, both now extinct.

In conjunction with the arrival of these northern species, many native South American species went extinct, especially many of the marsupials and the endemic hoofed mammals. In North America, the southern immigrants seemed to have little impact on the fauna. Paleontologists have suggested a number of hypotheses to explain this. One was that marsupials were intrinsically less able to compete with placental mammals, but this has no compelling biological basis, and it would not explain the extinction of the endemic placental hoofed mammals. The more likely explanation is ecological, and not unlike the explanation for island biodiversity. South America was smaller and had been isolated from other continents, whereas North America was larger and frequently connected to Europe, Asia, or both. Thus, North America had a greater pool of species with large populations than South America. North America is also mostly temperate, whereas the bulk of South America's landmass is in the tropics. One idea is that it was easier for temperate North American taxa to adapt to the tropical habitats of South America than for tropical South American taxa to adapt to temperate North American habitats.

Further Reading

Benton, M. J. 2014. *Vertebrate Paleontology*. Wiley-Blackwell, Malden, MA.

CHAPTER 8 Investigating the Origin of Life

Darwin did not actually address how life originated, only how it evolved after it originated. Darwin's evolutionary ideas require life to exist already. So, how did life begin? To answer that, we need to consider a transition from nonliving (or **abiotic**) chemistry to the chemistry of life. To do that, we must first consider what that transition entails. As it turns out, there are evolutionary aspects to this process, as well.

8.1 Qualities of life

What distinguishes the living world from the abiotic world? There are a number of qualities that are often ascribed to life, including homeostasis, metabolism, reproduction, heredity, organization (with cells as the fundamental unit), growth, adaptation, and irritability (i.e., ability to sense and respond to stimuli). While these are all things we associate with life, not all living things exhibit all of them: some unicellular organisms do not necessarily grow. Other qualities are really means to achieving another quality: cells create isolated units that can maintain homeostasis, and heredity is critical to reproduction.

For our purposes, we will consider three qualities that seem to be common to all living things. One is **metabolism**. All living things can transform chemical energy (and in some cases other types of energy) into other types of chemical energy. Cellular processes can transform energy to make cellular components (**anabolism**) or release energy by decomposing organic molecules (**catabolism**). Life is sometimes defined as "the controlled release of energy" to reflect the importance of metabolism.

Homeostasis refers to maintaining a constant internal environment, which could mean the inside of a cell or the inside of a multicellular organism. The chemistry of metabolism needs specific, stable conditions, so homeostasis helps to maintain metabolism, and, at the same time, metabolic processes actively maintain homeostasis.

Life exists because it persists, and that persistence is due to two things. One is the ability of individual organisms to maintain homeostasis over time. The other is an organism's ability to create more of itself, thus making it less likely that its kind will disappear. **Reproduction** is the process of creating more living individuals, and it is something that all living things do.

These qualities could constitute how we determine whether a system qualifies as being alive. For instance, how could we identify life on another planet where the conditions would not lead to the kinds of carbon-based life on earth? How could we identify life based on, say, silicon? Based on what we have discussed, we would expect any such life to be able to transform energy, maintain an internal environment, and make more of itself.

Another way to think of life is in terms of its history. As we discussed in Chapter 5, life is a monophyletic group. The qualities and processes that are common to all life presumably go back to its common ancestor, and life has persisted by changing and diversifying, evolving the many kinds of organisms that exist now and that existed in the past. One thing to keep in mind, in this view, is that the first living things were not necessarily the most recent common ancestor of all living things. First of all, it is possible that there were multiple origins of life, but that all of these lineages ultimately went extinct, except one; that one was able to persist and outcompete the others and eventually gave rise to all living things we know today and from the fossil record. Even that one lineage might have given rise to multiple other lineages through multiplication of species, but only one of those lineages persisted to become the ancestor of all life today. You sometimes see references to **LUCA**—the **last universal common ancestor**. LUCA is what is represented by the node at the very base of the tree of life. But LUCA is not the same as the first living thing, because LUCA is likely a descendant of some earlier organisms that resulted from the origin of life, and there might even have been multiple origins of life, only one of which—and not necessarily the first—gave rise to LUCA.

8.1.1 Viruses

You might be wondering how viruses fit into all of this. Viruses consist of DNA or RNA and some protein, such as a protective protein coat. Viruses enter cells and use the cellular machinery to replicate their nucleic acids and translate them into their proteins. Thus, viruses do not, as individuals, exhibit metabolism or homeostasis—rather, they parasitize those features of cells to reproduce themselves. Viruses therefore do not exhibit all of the qualities that we associate with life.

So, where did viruses come from? Do they represent some early stage in the origin of life? That is a possibility, but it is just as—if not more—likely that viruses are rogue bits of cells that have taken up this parasitic existence, or even former cells that have adapted to parasitizing other cells, losing along the way their qualities that would allow them to persist as individual cells.

8.2 The nature of the cell membrane

There are basically three kinds of components that are essential for cell-based life. First, it needs catalysts for the reactions of metabolism; these are generally enzymes made of protein. Second, it needs some kind of heritable material that can encode the information for building proteins and that can be replicated for new individuals resulting from reproduction—namely, nucleic acids. Finally, it needs a way of compartmentalizing these processes so that their chemistry is isolated from the environment. This is accomplished with a cell membrane made of **lipids**, molecules that make up fats and oils.

Lipids are generally **nonpolar**, as opposed to **polar**. Polar molecules have an uneven distribution of charges, due to how their electrons behave. Water molecules, for instance, consist of two hydrogen atoms and one oxygen atom. Oxygen has a high **electron affinity**, so it tends to attract electrons, such that the oxygen portion of water is more negatively charged, whereas the hydrogen portions are more positively charged. Nonpolar molecules have a relatively even distribution of charge across the molecule, so there is no part that is more positively or negatively charged.

Polar and nonpolar molecules interact in particular ways. If you are making a solution, the rule is "like dissolves like": polar molecules dissolve more readily in solvents made of polar molecules, whereas nonpolar molecules dissolve more readily in solvents that are nonpolar. Water is polar, so substances that are polar, like salt and sugar, will dissolve easily in it, whereas these same molecules will not dissolve in nonpolar solvents like vegetable oil or gasoline. This is also why oil and vinegar in salad dressings will always separate: oil is nonpolar and will not dissolve in water-based vinegar. Note that this does not mean that no oil dissolves in water or vice versa; we say that oil has a low **solubility** in water, because only a miniscule amount of oil will dissolve in the water.

Because water is the dominant solvent in the chemistry of life, we often describe polar and nonpolar substances in terms of their affinity for water. Polar molecules are **hydrophilic** (Greek for "water loving"), they will readily align themselves to be next to water. Nonpolar molecules are **hydrophobic** ("water fearing") and will align themselves to be as far from water as possible. Even when you shake up your bottle of salad dressing, the hydrophobic oil will move to be as far from the water as possible, forming a separate layer.

The nature of hydrophilic and hydrophobic molecules is important for understanding how the cell membrane can form compartments in what are generally watery environments. Our cell membranes are made of lipids called **phospholipids** (Figure 8.1). Phospholipids are molecules that have both polar and nonpolar parts. Each phospholipid has a head, made of a **phosphate** and **glycerol**, and a pair of parallel **fatty acid** tails. The phosphates and glycerol

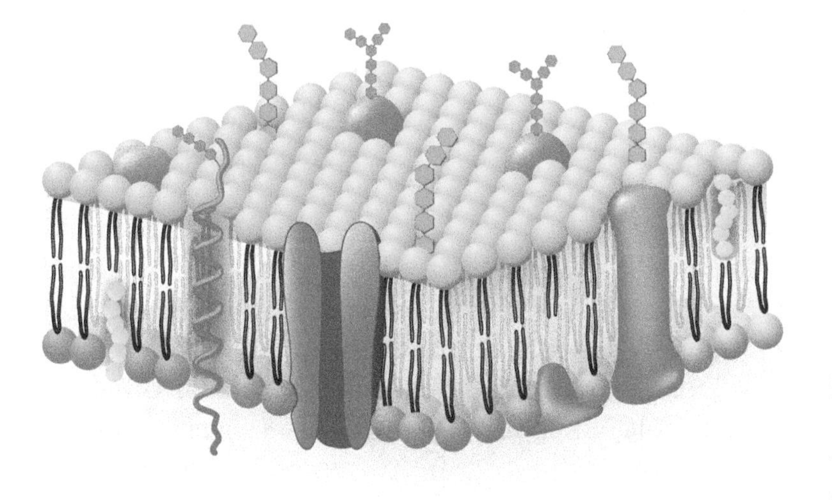

Designua/Shutterstock.com

Figure 8.1. Structure of the cell membrane. Phospholipids comprise most of the membrane and are represented by the yellow spheres (the phosphate–glycerol heads) on the inner and outer surfaces and the dark brown tails in the middle of the membrane. The blue, green, and purple structures represent proteins and polysaccharides embedded in or attached to the membrane.

make the heads of phospholipids hydrophilic. The fatty acids are hydrophobic. Thus, the head of a phospholipid will preferentially align with water, whereas the tail will be repelled by water. So, if you formed a single layer of phospholipids in water, their preferred arrangement would be with the heads against the water and the tails facing away from the water.

The phospholipids of a cell membrane are arranged in a **phospholipid bilayer**. This means that there are two layers of phospholipids, with the heads on the two surfaces of the membrane and the tails facing each other on the inside of the membrane. The membrane has an outer surface, facing the extracellular environment, and an inner surface facing the inside of the cell. Because both the extracellular environment and the medium inside of the cell—the **cytoplasm**—are based on water, this arrangement places the heads near the water and the tails away from the water.

The cell membrane also includes various proteins and other molecules with important roles, such as forming channels for molecules to cross the membrane or for binding to signaling molecules inside and outside of the cell, but the phospholipid bilayer is what makes the inside of the cell a separate environment from the outside. To cross the phospholipid bilayer itself, a molecule would have to be able to interact with the polar phosphate-glycerol heads and the nonpolar fatty acid tails—it would have to be a molecule that could navigate both the vinegar and the oil in a salad dressing. Very few molecules have this property, so the phospholipid bilayer is an effective barrier between the cell and the environment.

8.3 The origins of biological molecules

What we established above is that there are three key types of biological molecules—proteins, nucleic acids, and phospholipids—that contribute to each of the three major functions for a living cell—metabolism, homeostasis, and reproduction. Specifically, proteins catalyze reactions, nucleic acids encode replicable, heritable information, and phospholipids provide **compartmentalization**, sequestering a part of the environment that the cell can manage to suit its needs. The next question is: How did these biological molecules first arise?

Since we cannot go back in time to observe the origin of life, and because this event has essentially no chance of leaving a fossil record, the approach to understanding life's origins is generally to test hypotheses of how it could have happened based on what we know about the early earth. The conditions on the early, lifeless earth were very different from what they are today. Oxygen was essentially absent from the atmosphere. Oxygen is very reactive, and any oxygen that was produced or released through geological or other abiotic processes would react with other molecules in the environment. For instance, any iron exposed to the atmosphere would react with oxygen to form rust or iron oxide. It was only after photosynthesis evolved that oxygen started to accumulate in the atmosphere, because oxygen is a product of photosynthesis. For the purposes of the origin of life, the absence of oxygen is actually a benefit, as early biological molecules might have reacted with oxygen and would essentially have been lost.

The early earth was also much more energetic than the earth is today, at least at the surface. Heat and lightning would have been abundant from tectonic activity and the stormy atmosphere. The absence of atmospheric oxygen also means that there was no ozone layer, and this would allow ultraviolet (UV) radiation to reach the earth far more often than it does today. UV radiation is high in energy and can drive chemical reactions.

Before life existed, the chemistry of the earth would have included abiotic sources of the elements that make up living organisms. There of course would have been abundant water as the oceans formed. Nitrogen would have been available in the atmosphere not only as nitrogen gas (N_2) but also as ammonia (NH_3). The atmosphere and the oceans would also have contained other simple gases and liquids, like methane (CH_3), cyanide (HCN), and hydrogen sulfide (H_2S). At this time in earth history, extraterrestrial objects still bombarded the planet, and these often contain organic compounds that would have been introduced to the earth's chemistry. So, from these simple chemicals, how could complex biological molecules have formed?

The most famous experiment addressing this question is the Urey-Miller experiment, first published in 1953. Harold Urey was the advisor of Stanley Miller, then a graduate student, and it was Miller who designed and executed the experiment (Figure 8.2). Miller set up a series of flasks and tubes as a closed system: in other words, the experiment only contained what he put into it, as Miller used a vacuum to empty it of any other molecules. He then introduced various simple gases that would have been present in the atmosphere of the early earth: methane, ammonia, and so on. He introduced energy in the form of an electrical spark. He let the experiment run and sampled the fluid that collected from the reactions it produced. The products included amino acids, demonstrating that amino acids could be produced by abiotic processes on the early earth.

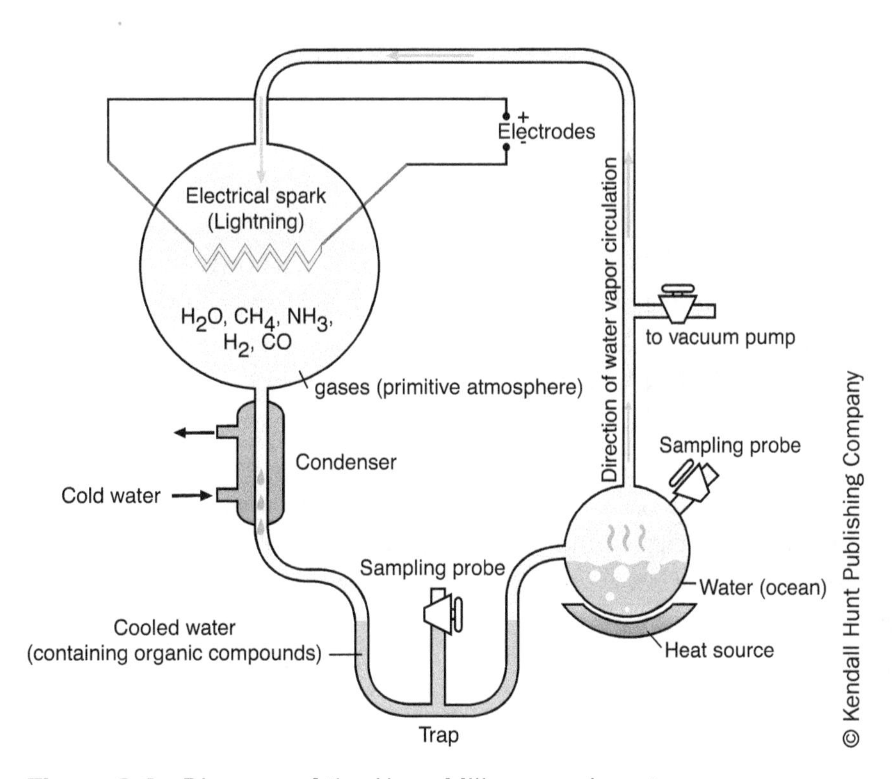

Figure 8.2. Diagram of the Urey–Miller experiment.

Miller continued to run this experiment throughout his life, and others replicated it and modified it to use other kinds of energy, such as heat and UV light. The results produced more kinds of amino acids and even some nucleic acids. Thus, Miller's experiment demonstrated how the basic components of two of the three most important molecules for life could form in the absence of life.

8.4 Getting the processes of life started

Once biological molecules were available, how did they organize themselves into the processes that support life? There is a particular "chicken and egg" problem that arises from this question. Proteins are constructed in living organisms today through the transcription and translation of nucleic acid sequences. But transcription and translation—not to mention replication of nucleic acids—require specific enzymes, which are proteins. Which came first, and how could one function without the other?

The current answer to this comes from studies of RNA. We tend to think of DNA when we think of the important functions of nucleic acids, but RNA has many of the same properties of DNA that would be important for the basic functions of life. RNA also has been shown to have a property that DNA does not: there are RNA sequences that can effectively act as **polymerases**, which is the class of enzymes that facilitate copying nucleic acids, whether for DNA replication or for transcribing DNA to RNA. These catalytic RNA sequences are called **ribozymes**. Thus, RNA could have replicated itself and could have catalyzed the processes of translating itself into protein. Evolution of RNA could lead to producing more complex proteins that take over the polymerase role, eventually introducing the more stable molecule DNA as the primary repository of information. This early, RNA-dominated stage of life's evolution is referred to as the "**RNA world**."

But, what about the membranes? Lipids are less complex than proteins and nucleic acids and could either have formed spontaneously in the chemistry of the early earth or have been introduced by extraterrestrial objects. Lipids are also known to spontaneously form **vesicles**, little lipid-bound pockets, which could have provided early compartmentalization for ribozymes and other RNA sequences (Figure 8.3). Some ribozymes can synthesize phospholipids, and the addition of phospholipids to simple lipid vesicles makes them more permeable, which would allow an early RNA-based cell to exchange molecules with the environment, taking in nutrients and expelling waste.

The origin and early evolution of life is still an active area of research, and there are many questions still to be answered. The processes described here might have occurred multiple times, only for the vast majority to fail at some point, or for one instance to succeed to the point of outcompeting the others. Only later, and perhaps after numerous branches from this lineage had diverged and gone extinct, would one lineage survive, either by chance or by virtue of its advantages, to form the cells that would lead to LUCA.

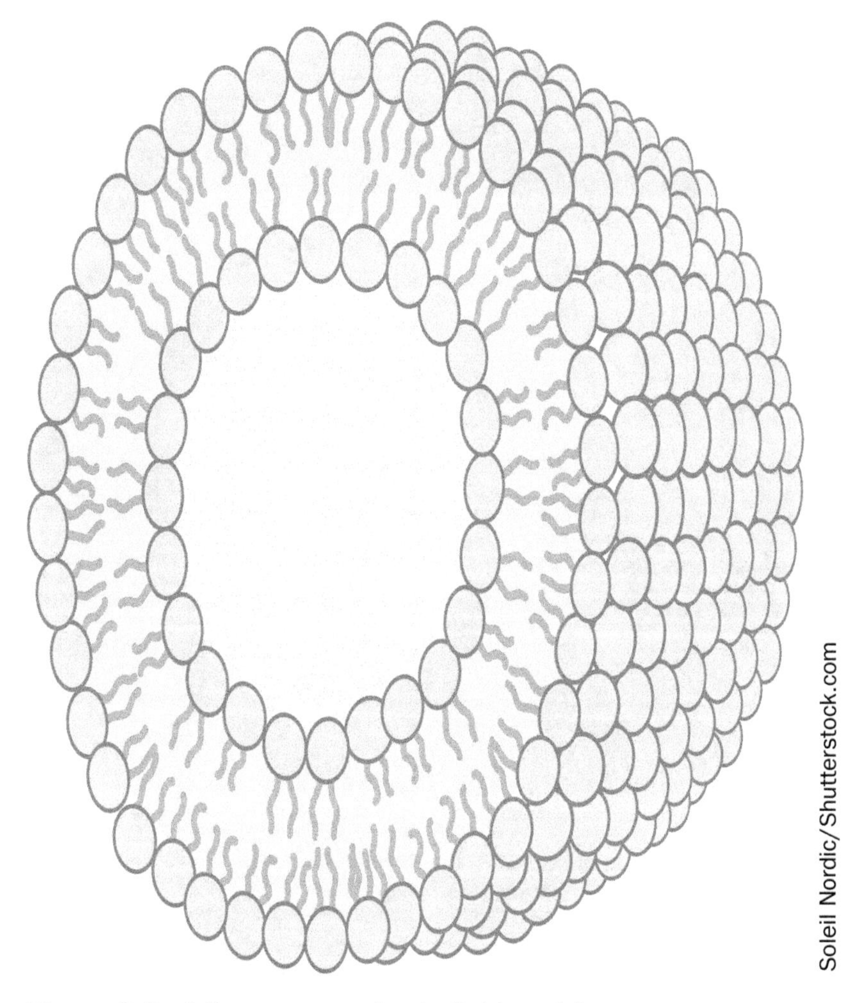

Soleil Nordic/Shutterstock.com

Figure 8.3. A liposome, a simple lipid vesicle.

Further Reading

Chen, I. A., and P. Walde. 2010. "From Self-Assembled Vesicles to Protocells." *Cold Spring Harbor Perspectives in Biology* 2(7): a002170. doi:10.1101/cshperspect.a002170.

Yarus, M. 2011. *Life from an RNA World: The Ancestor Within*. Harvard University Press, Cambridge, MA.

CHAPTER 9

An Appallingly Brief Survey of Biodiversity

The focus of this book so far has been on broad, fundamental concepts concerning how evolution works rather than on specific details of what it has produced. Nevertheless, evolution has produced many very specific details in the unique combinations of inherited traits and adaptations that distinguish each of the many millions of species on this planet. In the course of this book, we have already been introduced to a large number of species from disparate parts of the tree of life. Thus, it is also important for a student of biology to have some sense of the pattern of biodiversity in order to provide some context for all of these broad principles. This chapter aims to provide some of that context through a survey of biodiversity that is of necessity exceedingly shallow. The goal here is to provide a scaffold upon which students might later support more detailed studies of particular groups.

9.1 Prokaryotes and eukaryotes

A fundamental division of living organisms is into what are called **prokaryotes** and **eukaryotes**. As we will discuss, the ancestor of all life today was presumably a prokaryote, whereas eukaryotes are a single but very diverse lineage descended from prokaryotes. Because both kinds of organisms include at least some unicellular forms, the fundamental differences between them are at the subcellular level. There are actually a number of ways in which prokaryotes and eukaryotes differ, but the main one, or at least the most familiar one, is that eukaryotes have a **nucleus** and prokaryotes do not (Figure 9.1). To put it another way, eukaryotes surround the DNA that comprises their genome with a **nuclear envelope**, an internal membrane similar to the cell membrane that we discussed in the Chapter 8. The DNA of prokaryotes floats freely in the cytoplasm.

A more general difference between eukaryotes and prokaryotes is that eukaryotes have many internal membranes enclosing other structures or functional spaces. Internal cellular structures with specific functions are called **organelles**, and **membrane-bound organelles** are unique to eukaryotes. One function these organelles could perform is dealing with the increasingly oxygenated atmosphere of the earth starting over two billion years ago, and this might have been a driver of eukaryotic success. Eukaryotic plasma membranes tend to be more extensive and flexible than they are in prokaryotes, and eukaryote cells are typically larger than those of prokaryotes. Eukaryotes and prokaryotes also

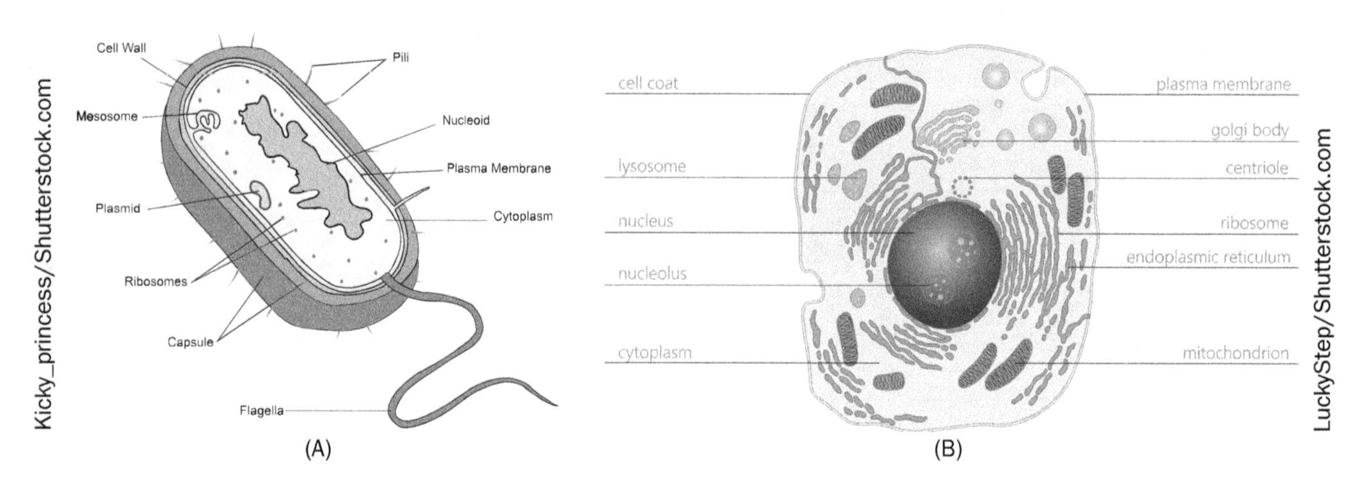

Figure 9.1. Prokaryotes versus eukaryotes.

differ in how their DNA is organized: prokaryote DNA tends to be contained in a single, circular chromosome, whereas eukaryote DNA, besides being confined to the nucleus, is typically organized into multiple chromosomes and wound around special proteins called **histones**.

9.2 The three domains

We mentioned in Chapter 5 that the Domain was introduced in the twentieth century as a rank higher than Linnaeus's highest rank of Kingdom. By the 1970s, our understanding of the microbial world had made Linnaeus's two Kingdoms of Plantae (plants) and Animalia (animals) inadequate. It was clear that life could be divided into more kingdoms, and for a while five kingdoms were recognized: in addition to plants and animals, fungi were given their own kingdom, and unicellular microorganisms were divided into the prokaryotic Kingdom Monera and the eukaryotic Kingdom Protista.

But two things made even this arrangement problematic. First, it was increasingly clear that certain eukaryotic microbes were actually more closely related to animals, plants, or fungi than they were to other eukaryote microbes, so Protista was not monophyletic. Furthermore, it was discovered that some prokaryotes were distinct from other more familiar prokaryotes, like bacteria. These distinct prokaryotes were called **archaebacteria**, and ultimately this led to the recognition of three domains of life: **Domain Bacteria**, including the more familiar and abundant prokaryotes that we know as bacteria; **Domain Archaea**, including the archaebacteria, or **archaeans**; and **Domain Eukarya**, including all eukaryotic organisms (Figure 9.2).

Within each of these domains are one or more kingdoms, though as our understanding of the phylogeny within each domain progresses, we find more complexity than can easily be captured by binning things into a small number of kingdoms; this, as we shall see, is particularly true for Eukarya.

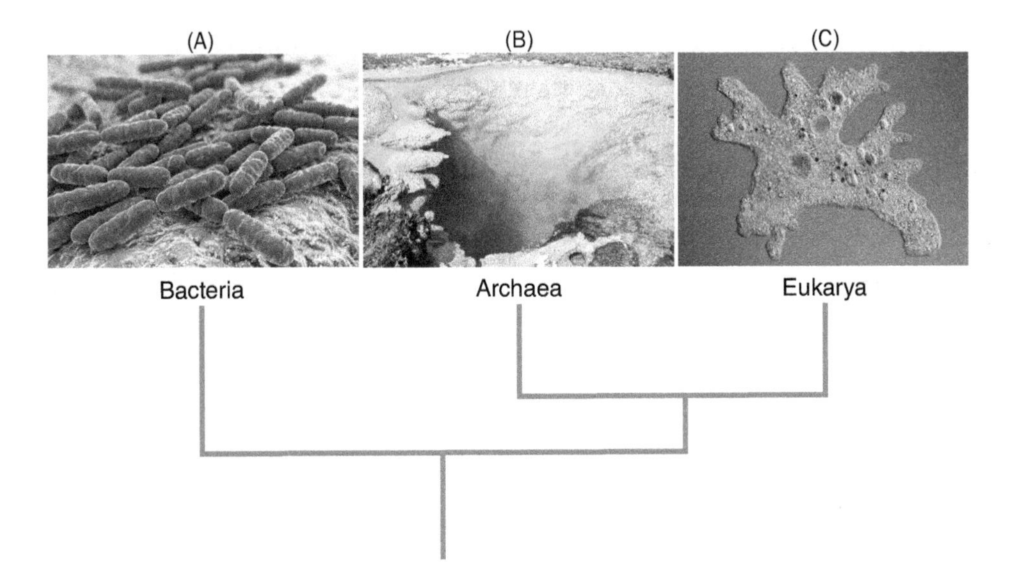

Figure 9.2. The three domains.
9.2.A Kateryna Kon/Shutterstock.com
9.2.B James Mattil/Shutterstock.com
9.2.C Lebendkulturen.de/Shutterstock.com

9.2.1 Bacteria

The diversity of bacteria is far too large to include in the scope of this book, but it is worth mentioning a few things about them. First, it should be noted that the diversity of bacteria (and, for that matter, archaeans) is not so much about what they look like but what they do at the biochemical level. Thus, bacteria are often adapted to specific kinds of chemical and physical conditions, like environments with a particular pH, temperature, or salinity, or environments with an abundance of iron or some other chemical. We often think of pathogens when we think of bacteria, but bacteria do much that enhances our lives, including many things that other organisms cannot do. Some bacteria are important for nitrogen fixation, which means they can incorporate nitrogen from the atmosphere into biological molecules. This is a vital function for life because many important biological molecules, including all proteins and nucleic acids, contain nitrogen atoms. Some other bacteria can digest the cellulose that makes up plant walls and that cannot be digested by most herbivorous animals. Bacteria living in the guts of certain herbivorous mammals make the nutrients from cellulose available to these animals. This mutually beneficial relationship between the herbivore and the bacteria is called **symbiosis**.

Some bacteria are **photosynthetic**, meaning they can produce sugars from water and carbon dioxide using energy from sunlight. The most important photosynthetic bacteria are **cyanobacteria**, and the first photosynthesis that happened on earth was likely from a cyanobacterium or something like it. Cyanobacteria are responsible for the stromatolites we discussed in Chapter 7.

Photosynthetic organisms are **autotrophs**, because they produce nutrition using energy from their environment, rather than consuming organic molecules like **heterotrophs** do.

Because they use light, they are called **photoautotrophs**. But photosynthesis is not the only way to be an autotroph. Some bacteria live deep in the ocean along cracks in the ocean floor that erupt with heat and chemicals from beneath the earth's crust. These environments are called **hydrothermal vents**, and they can reach temperatures up to 400°C; only the great pressure at those depths keeps the water from boiling. The bacteria that live on these vents are so far from the surface that they cannot use sunlight for photosynthesis. Instead, these bacteria are **chemoautotrophs**, using energy from inorganic chemical reactions to power the production of their nutrition.

9.2.2 Archaea

Archaea were originally thought to be associated with extreme environments, namely high temperature environments, including hydrothermal vents, or high salinity environments. It is now known that archaeans are found in a wide variety of environments, including in the human gut. The only organisms that are **methanogens**, meaning they can produce methane as part of their metabolism, are archaeans. Analysis of DNA sequences has shown that Archaea is more closely related to Eukarya than it is to Bacteria (Figure 9.2). Archaeans also share with eukaryotes the presence of histones.

9.3 Eukaryote diversity

Eukaryotes include a wide variety of organisms, including all of the familiar multicellular groups that we typically associate with life, namely animals, plants, and fungi. There are also a host of unicellular eukaryotes, some of which are familiar subjects of biology classes (Figure 9.3). These microbes are frequently put into groups like "protozoans" (for the more animal-like ones) or "algae" (for the more plant-like ones), but these terms disguise a much more complex set of relationships.

The origin of eukaryotes is an interesting topic, because it appears that at least some of the membrane-bound organelles in a eukaryotic cell are descended from free-living prokaryotes that somehow were incorporated into another cell. The best example of this comes from **mitochondria**, organelles that are responsible for the metabolic process of aerobic respiration that allows eukaryotic cells to produce an abundance of chemical energy from sugars. Mitochondria are structurally similar to some prokaryotes, but what is most telling is that mitochondria have their own DNA, and that the DNA of mitochondria is most similar to that of certain bacteria. **Chloroplasts**, the sites of photosynthesis in plants and certain algae, also have their own DNA and appear to be descended from certain cyanobacteria.

These discoveries prompted the theory that eukaryotes originated through a process of **endosymbiosis**. What this means is that the various organelles of eukaryotic cells, especially the membrane-bound ones, are descended from free-living prokaryotes that were somehow engulfed by a larger prokaryote, but that the engulfer and the engulfed persisted and coevolved to act as a single unicellular organism. In other words, the organelles were once symbiotic prokaryotes living inside of a host cell.

Figure 9.3. Eukaryote microbes. (A) *Paramecium*. (B) *Amoeba*. (C) *Volvox*. (D) Kelp. (E) Green algae. *Volvox*, kelp, and green algae are colonial algae. (F) A diatom.

9.4 Multicellularity and the major multicellular eukaryote kingdoms

Eukaryotes include essentially all of the groups that we consider to be **multicellular**. By this, we mean that these organisms are made of many adjoined cells, that their cells are specialized for particular roles, and that their cells depend on these different roles for their survival and reproduction. Multicellular organisms are distinguished from colonial organisms, which are collections of individual cells that are physically joined together but where each cell otherwise acts as an independent organism. In fact, these kinds of distinctions might better reflect points on a spectrum. There are even some organisms, like slime molds, that alternate between living as unicellular individuals and as multicellular organisms.

There are three major kingdoms of multicellular eukaryotes: **Fungi**, **Plantae** (or **Metaphyta**), and **Animalia** (or **Metazoa**). The closest relatives of each of these kingdoms are different kinds of unicellular eukaryotes, indicating that multicellularity arose independently in each kingdom. Furthermore, while fungi were for a long time treated as "plants," they are not only distinct from plants but actually more closely related to animals than to plants (Figure 9.4).

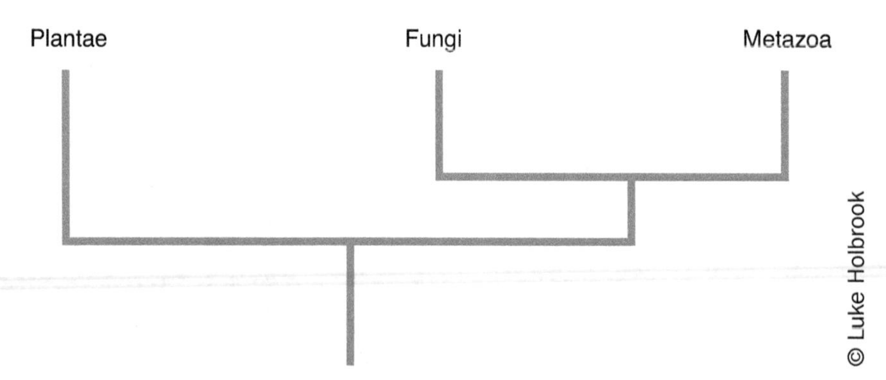

Figure 9.4. Relations among multicellular phyla.

9.4.1 Fungi

Fungi are usually multicellular organisms that consist of filaments, called **hyphae**, and their cells have a cell wall made of **chitin**, a fibrous substance that also contributes to the exoskeletons of insects and other arthropods (Figure 9.5). There are some unicellular fungi, namely yeast. Fungi are **heterotrophs**, which means they do not produce their own nutrition but must consume organic molecules from the environment. Fungi generally absorb small organic molecules around them, like those resulting from decay of dead organisms, taking advantage of the abundant surface area that the hyphal filaments present. These filaments extend through a substrate and can in some cases become massive; the largest organism on earth is thought to be a fungus growing in and under over 3.7 square miles of forest in Oregon. However, we often don't notice fungi unless we go looking for them, or until they grow large enough to form a spot of mold on a piece of old bread. Many of us identify fungi with mushrooms, but mushrooms are actually reproductive structures formed by a larger individual.

We won't get into the diversity of fungi, but it should be noted that fungi are important in many aspects of our world. They directly benefit us commercially as leavening for bread and fermenters of alcohol, and they also have produced antibiotics, most famously penicillin. Some are agents of infection, and some spoil our food. Fungi are important decomposers in the environment, helping to release the biological molecules trapped in dead organic matter.

Irina Kozorog/Shutterstock.com

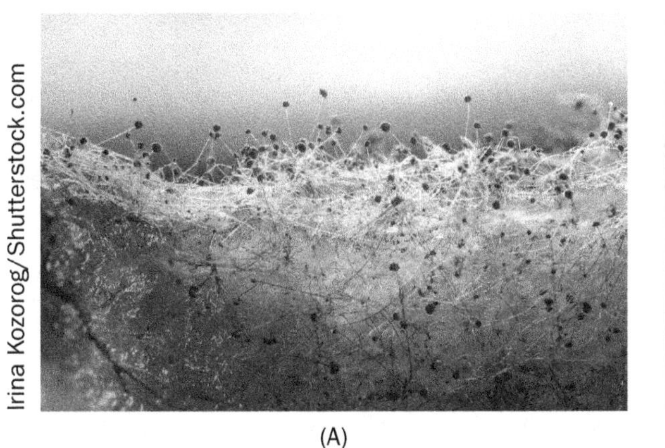

(A)

(B)

ressormat/Shutterstock.com

Figure 9.5. Fungi. (A) Bread mold. (B) Mushrooms.

Some fungi, known as **mycorrhizae**, are associated with the roots of many plants, and without this association the plants cannot survive in certain environments. Thus, mycorrhizae are critical in many ecosystems.

9.5 Plants

Plants form the energetic base of almost all terrestrial environments, using energy from the sun to convert water and carbon dioxide into sugars, through the process of **photosynthesis**. But photosynthesis appeared billions of years before plants arose, and plants have only been around since they started colonizing the land in the Ordovician. In fact, most photosynthesis occurs in the oceans, mainly performed by various marine algae known as phytoplankton. In plants, photosynthesis is largely facilitated by organelles called **chloroplasts**, which contain a pigment called **chlorophyll** that absorbs the light that powers photosynthesis.

Plants are distinguished from other photosynthetic organisms by their adaptations to life on land. Their cells have rigid walls of **cellulose** that help them to remain upright. They are covered by a waxy **cuticle** that minimizes water loss. The undersides of leaves, where most photosynthesis occurs in a plant, have openings called **stomata** (s. stoma) that allow for gas exchange and that can be closed during the day to prevent water loss (Figure 9.6). Plants typically have **roots** that anchor them in the substrate and provide surface area to absorb nutrients from the soil.

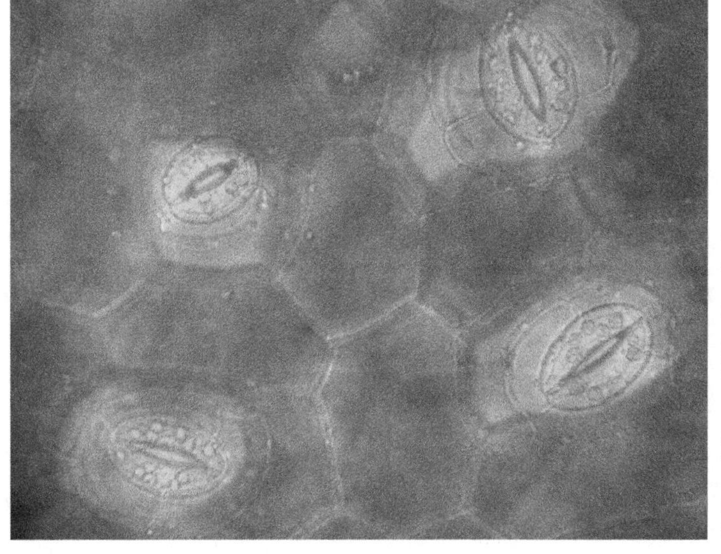

Kollawat Somsri/Shutterstock.com

Figure 9.6. The underside of a leaf showing stomata.

9.5.1 Alternation of generations

A distinguishing feature of the life history of plants is that they alternate between two distinctly different types of individuals, a pattern called **alternation of generations** (Figure 9.7). In the case of plants, they alternate between **haploid** individuals called **gametophytes** and **diploid** individuals called **sporophytes**. Haploid individuals have half the number of chromosomes of diploid individuals, because diploid individuals have chromosomes from both parents, meaning their chromosomes are in sets of **homologous pairs**, whereas haploid individuals only have one chromosome from each set, potentially coming from either parent.

As the name suggests, haploid gametophytes produce **gametes**, haploid sex cells that come in two forms, eggs and sperm. When an egg is fertilized by a sperm, a diploid cell is formed that will divide to form a multicellular sporophyte. In order for the sporophyte to complete the cycle, it needs to form **spores**, individual haploid cells, and it does so through a process called **meiosis**, whereby a diploid cell gives rise to haploid daughter cells. The spores are often dispersed and later divide to form multicellular gametophytes.

It is usually not obvious to us that plants come in these two forms, but, as we will see, all plants exhibit this life cycle, and specific lineages of plants differ in the relationship between these different generations.

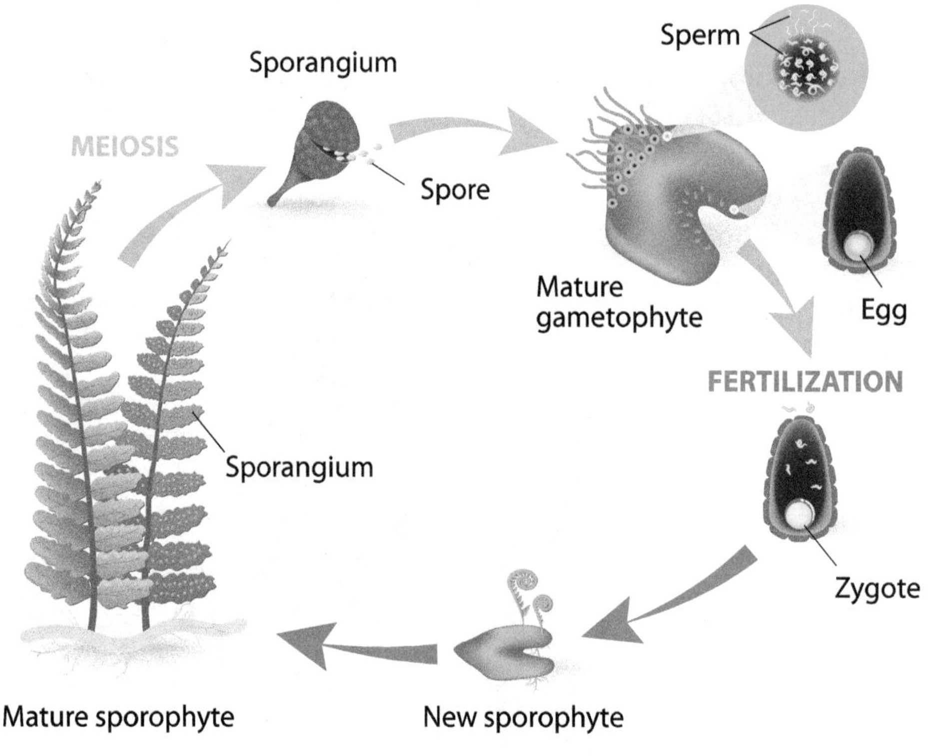

Figure 9.7. Alternation of generations illustrated for a fern.

A Closer Look: Mitosis and Meiosis

A key part of how multicellular organisms function is how their cells divide to form new cells. A key part of this process is ensuring that each cell gets the genetic material that it needs to function. Most of your cells carry the same diploid set of chromosomes, so forming a new cell needs to include copying the genetic material and apportioning it to the **daughter cells**, which are the two new cells resulting from a single cell dividing. We have encountered how the genetic material is copied in Chapter 5 when we discussed replication of DNA. The process of moving those two copies of diploid chromosomes into the forming daughter cells is called **mitosis**.

When a cell is not dividing, its nuclear DNA is diffused throughout the nucleus as threads of **chromatin**. When the cell is ready to divide, the chromatin condenses into paired chromosomes, where the DNA is densely packed. When the DNA replicates, which actually happens before mitosis, the chromosomes are doubled, with the copies connected at a point called the **centromere**. Long thin proteins called **spindle fibers** extend from a point near these centromeres to structures on opposite poles of the cell called **centrosomes** (each made of two parts called **centrioles**). Essentially, the spindle fibers pull the chromosomes apart at the centrosome and toward the respective poles, such that one copy of each chromosome is moved to each pole (Figure 9.8).

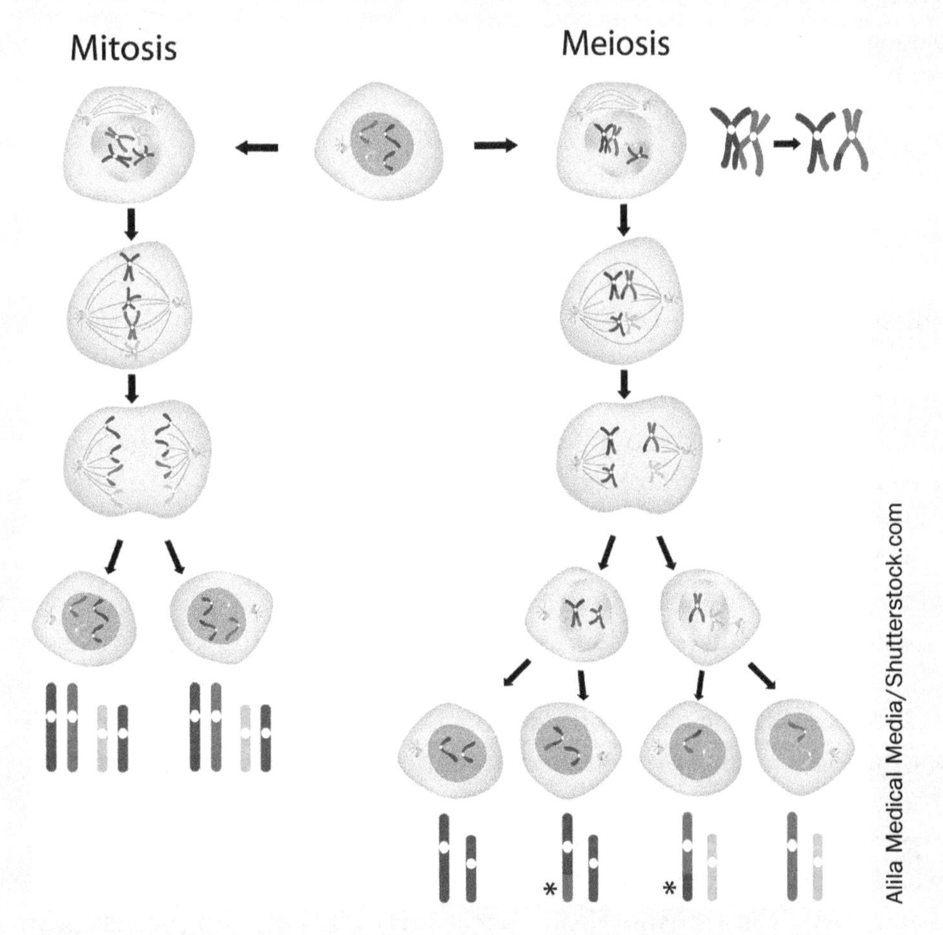

Figure 9.8. Mitosis and meiosis. The four stages illustrated for mitosis correspond to (from top to bottom) prophase, metaphase, anaphase, and telophase.

The process is described as having four stages: **prophase**, when the chromosomes are condensing; **metaphase**, when the chromosomes line up at the "equator" of the cell; **anaphase**, when the chromosomes are visibly separated toward the two poles; and **telophase**, when the chromosomes reach their respective positions and actual cell division happens, separating the cytoplasm into two new cells. Note that at the end, each cell should have the exact same chromosomes as the other.

Meiosis is the process of forming haploid daughter cells from a diploid cell, and therefore the process requires that at some point the paired homologous chromosomes are segregated into different cells. Like mitosis, meiosis happens after DNA replication, so each pair of **homologous chromosomes** (i.e., the corresponding chromosomes that were received from each parent) is doubled. The process is similar to that of mitosis, but it occurs in two sets of divisions, and the phases have the same names, except with "I" or "II" appended, to denote whether it is a phase of the first division or the second. The first division, **meiosis I**, separates the homologous chromosomes, such that each daughter cell is receiving the two copies of one chromosome from each homologous pair. Note that which chromosome—the mother's or the father's—that moves to a particular side is essentially random, such that a daughter cell is likely to receive a random assortment of the mother's chromosomes and the father's chromosomes.

The two daughter cells from meiosis I go through a second division, **meiosis II**, that is essentially analogous to mitosis, such that one copy of each chromosome goes to each new daughter cell. The end result is four haploid daughter cells.

9.5.2 Plant diversity and phylogeny

Figure 9.9 illustrates some major lineages of plants and their relationships. The plants that are probably most similar to the earliest plants include several lineages collectively called **bryophytes**. These include mosses, hornworts, and liverworts. It is likely that bryophytes are not a monophyletic group, but they are often discussed together for the ancestral plant features that they share. Bryophytes tend to be short plants, as they lack the rigidity to stand taller and lack specialized vessels to transport water and products of photosynthesis for long vertical distances between the leaves and the roots. They tend to live in moist environments, mainly because their sperm need to travel in water to fertilize their eggs. Finally, the gametophyte is the dominant part of the life cycle of bryophytes. The green, spongy matter that you see when you find a bed of moss is actually gametophytes. Sporophytes grow out of the gametophytes and often lack green pigment. Thus, they are physically connected to the gametophyte and depend on gametophyte photosynthesis for their nutrition.

Vascular plants, or **tracheophytes**, are a monophyletic group that differs from bryophytes in several important ways. Vascular plants have two types of specialized vessels—**xylem** and **phloem**—that can transport substances vertically between the leaves and the roots. Xylem transports water absorbed in the roots upwards to other parts of the plant, especially the leaves, where water is used in photosynthesis. Phloem transports sugars produced in

Figure 9.9. Plant diversity and phylogeny. (A) A clump of moss. Note the brown sporophytes projecting from the green gametophytes. (B) Fern sporophytes. (C) A branch of a Douglas fir, a conifer, with a cone. (D) The world's biggest flower, from the flowering plant *Rafflesia*.

Jeff Holcombe/Shutterstock.com
Settaphan Rummanee/Shutterstock.com
Neil Webster/Shutterstock.com
Alexander Mazurkevich/Shutterstock.com

the leaves by photosynthesis down to the roots. Vascular plants generally can grow taller than bryophytes, in part due to **lignin**, an organic substance that makes cell walls more rigid. Lignin is especially abundant in the walls of xylem cells and gives wood its distinctive rigidity and durability. The sporophyte is dominant in vascular plants: it is photosynthetic and nutritionally independent of the gametophyte, and it is typically much larger than the gametophyte.

The two major types of vascular plants that we will discuss here are **ferns** (or **pteridophytes**) and **seed plants** (or **spermophytes**). The long parallel-leaved fronds we typically envision when we think of a fern is actually the sporophyte. The gametophyte is smaller and lives as an independent individual. Like bryophytes, ferns rely on water for fertilization, and their spores are dispersed through the air.

Seed plants are a monophyletic group that includes the vast majority of what we normally think of when we think of plants. Unlike ferns, seed plants don't rely on water for fertilization. Sperm are packaged in **pollen**, protecting them from desiccation and allowing them to be dispersed through the air or by animal pollinators. They also have, as you might expect, **seeds**, which are embryonic diploid sporophytes packaged in a durable casing along with some nutrition in the form of a carbohydrate-rich **endosperm**. As in ferns, the sporophyte of seed plants is dominant, and in fact the gametophyte is nutritionally dependent on the sporophyte and even encased within it. The seed is actually composed of parts of three individuals: the embryonic new sporophyte, the gametophyte that produced the embryo, and the sporophyte that produced the gametophyte.

The two main groups of seed plants are **gymnosperms**, which include familiar plants like pine trees and other **conifers**, and **angiosperms**, or flowering plants, which are the most abundant and diverse plants by a wide margin. Both groups include species that grow into tall trees that make up forests familiar to us, though it should be noted that some ferns grow tall enough to be considered trees, and tree ferns were the main trees of the first forests on earth. Conifers produce cones that contain their reproductive structures. Flowering plants are distinguished by two important reproductive features of their own: **flowers** and **fruits** (Figure 9.10).

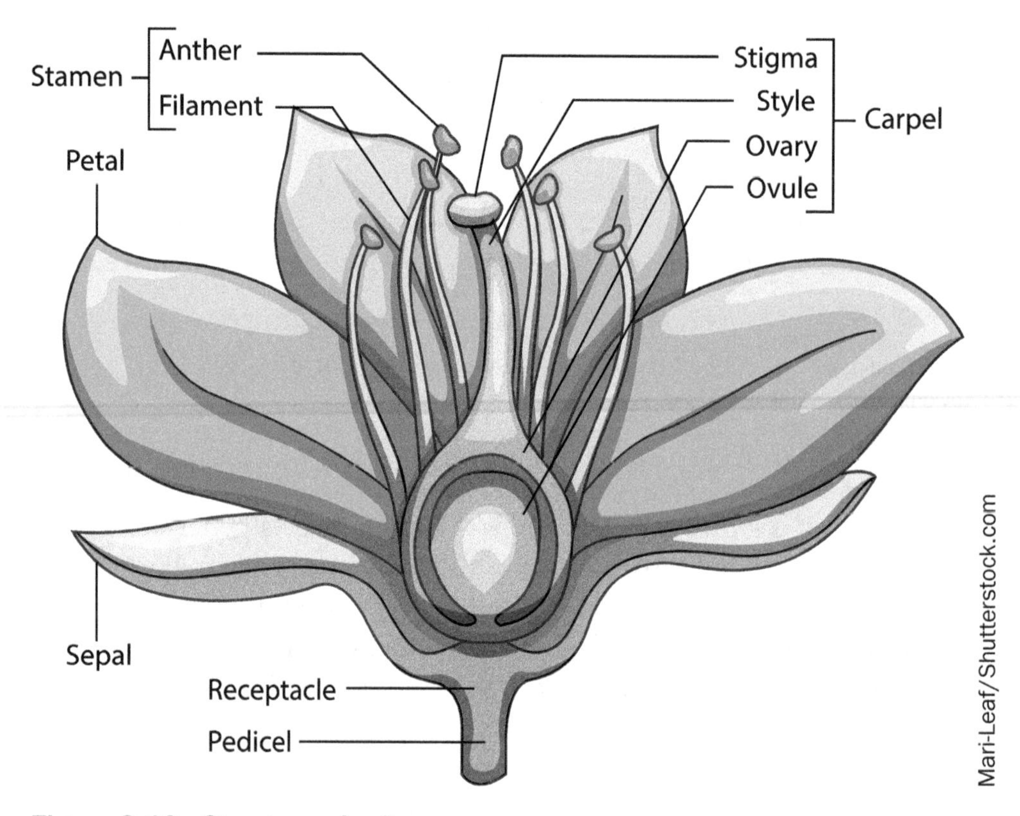

Figure 9.10. Structure of a flower.

Flowers are composed of several concentric layers of specialized leaves, including the **sepals** that surround the growing bud and the base of the flower, **petals**, **stamens** that contain the male pollen-producing structures, and, in the center, **carpels** that contain female structures, including the **ovary** that produces the egg. Flowers can be male or female, or they can bear structures of both sexes. Even if a plant species has flowers that are only male or female, a species might have individual plants that bear flowers of only one sex or individuals with flowers of both sexes.

Fertilization happens when a pollen grain sticks to the **stigma** at the top of the **style** protruding from the top of the carpel, and the pollen grain extends a tube through the style to the ovary that contains the **ovule**, which in turn contains the egg (or **ovum**). The ovary will then mature into a fruit, which is essentially a device for dispersing the seeds. We think of fruits as edible things, and edible fruits are essentially an enticement to animals to eat them, including their seeds, which will be dispersed in the animal's droppings elsewhere. But dispersal of seeds by means of fruits is not limited to ingestion by animals. Some fruits, like burrs, are not edible (or at least not appetizing), but attach themselves to animals that carry them away until they are brushed or scraped or fall off of the animal. The fuzzy stalks extending from dandelion seeds are part of a fruit that is adapted for wind dispersal. Then there are coconuts, whose thick but buoyant husks protect the seed as it floats in the ocean, until it washes up on a beach and can germinate.

Flowering plants dominate a wide variety of terrestrial environments. When we think of flowers, we tend to think of smaller plants with ornamental flowers, but flowering plants include most of the trees of temperate and tropical deciduous forests, the cacti of the desert southwest of the United States, and the grasses that cover the prairies and savannahs.

9.6 Metazoa and its major lineages

When you hear the term "animal," you probably first think of a furry mammal that is not a human. Maybe you are so broad-minded that your first thought is of a bird or a lizard or some other sort of terrestrial vertebrate. The upshot is that you tend to think of animals that are more like you—but still not humans—when in fact the biological meaning of "animal" encompasses a lot more, including fishes, insects, worms, and corals, among many others. And this meaning also includes humans.

Animals are a monophyletic group of organisms classified in the Kingdom Animalia, or more commonly in the **Kingdom Metazoa**. We will use "metazoan" instead of "animal" going forward, in part to avoid all the cognitive and cultural baggage that comes with the latter. Like fungi, metazoans are heterotrophs, though they usually consume larger particles that they break down into useable biological molecules.

Metazoans are multicellular organisms, and they differ from other multicellular eukaryotes in that their cells secrete substances outside of them that form an **extracellular matrix** (or **ECM**). The ECM can take many forms, but the cells are always embedded in, or stuck to, or floating around in this substance. The hard matrix of your bones, the plasma of your blood, and the basement membrane to which the cells lining the inside of your mouth adhere are all examples of ECM. All forms of ECM include proteins and/or polysaccharides. The most common protein in your body is **collagen**, which is an ECM protein and is unique to metazoans.

Metazoans also differ from other multicellular groups in how they develop. All multicellular organisms start as a single cell—a fertilized egg—and must go through some process of **development**, or **ontogeny**, where the single cell divides into many, interacting cells, and where those cells differentiate into different specialized cell types. All metazoans go through a stage of development known as the **blastula**. A blastula is a hollow ball of cells (Figure 9.11). Early in development, the fertilized egg divides into two, four, and eight cells, and so on, but it is still a solid ball of cells. At some point, a space forms in the middle of the ball,

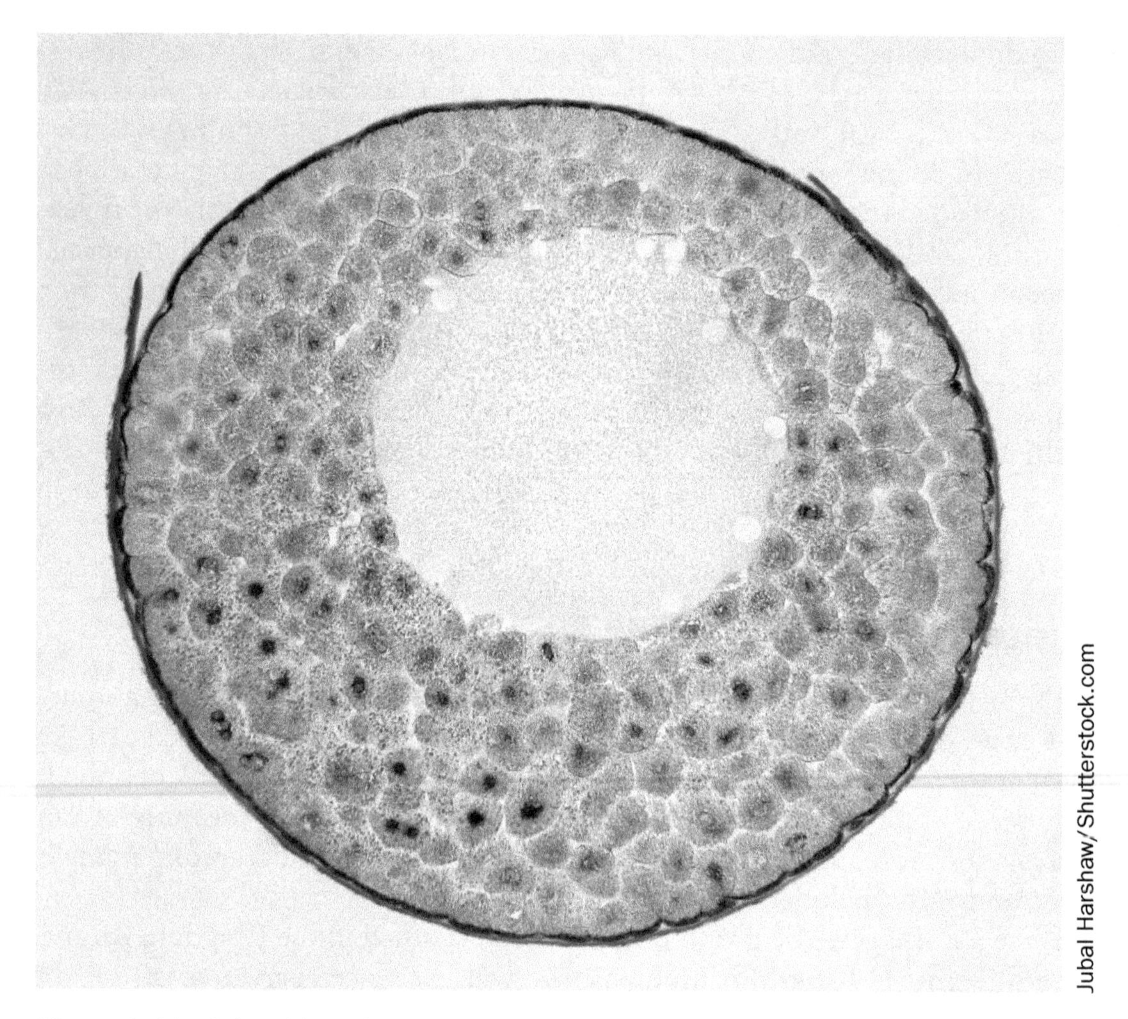

Figure 9.11. A frog blastula.

such that the cells now form the outer layer of a hollow sphere. The space inside is called the **blastocoel**.

After the blastula stage, development varies considerably among metazoans, and even the exact nature of the blastula varies among them. In fact, metazoan **embryology**, or the study of the patterns of development among metazoans, has been an important area of research for understanding metazoan diversity, and differences in development are often characteristic for major lineages.

Figure 9.12 shows a phylogeny of metazoans as we currently understand it. There are some controversies regarding even the relationships at the base of the metazoan tree, but Figure 9.12 reflects the current consensus and provides a starting point for examining questions of metazoan phylogeny elsewhere. Figure 9.12 also does not include every known phylum of metazoans but does include the ones that are most familiar and that figure most prominently in discussions of metazoan evolution. At the base of the metazoan tree, we can identify three main lineages: the **Phylum Porifera**, the **Phylum Cnidaria**, and the collection of phyla included in **Bilateria**.

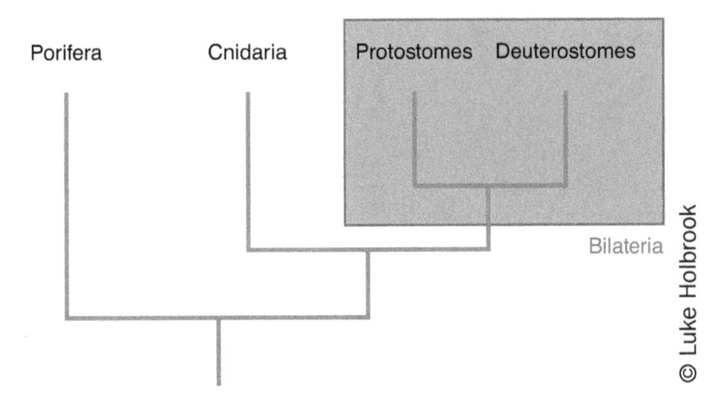

Figure 9.12. Metazoan phylogeny.

9.6.1 Porifera

Poriferans or **sponges** (Figure 9.13) are perhaps the simplest metazoans in structure. They are aquatic and are found mostly in marine environments, where they are **filter feeders**, sieving particles out of the water for their food. They have specialized cell types, but the cells are

Figure 9.13. Sponges of the Phylum Porifera. (A) A colony of calcareous sponges. (B) The skeleton of a glass sponge. (C) "Natural sponges"; these are the remains of sponges that have a skeleton made of the protein spongin.

not organized into **tissues**, as they are in other metazoans. The phylum name means "pore bearer" and refers to the openings, or pores, in the sponge's body. Water is drawn through these pores by the beating action of whip-like proteins called **flagella** on special cells called **choanocytes**, or collar cells. The mucous-covered collar around the flagellum collects organic particles that provide nutrition for the rest of the sponge. The water then enters a space in the middle of the sponge called the **spongocoel** and exits the sponge through a large opening called the **osculum**. Sponges are **sessile**, they attach to a substrate and do not move around.

Choanocytes of a sponge are very similar to some unicellular eukaryotes called **choanoflagellates**, and DNA evidence supports choanoflagellates as the closest living relatives of metazoans, indicating that metazoan multicellularity arose independently of that of plants and fungi.

Sponges do have a skeleton, but it is often made of tiny elements called **spicules**, which look like jacks from the old-fashioned game of the same name. Spicules can be made of calcium carbonate or glass, but in many sponges the skeleton is not made of spicules but of a protein called **spongin**. Sponges with this type of skeleton are the ones that are sold as "natural" sponges for washing.

9.6.2 Cnidaria and Ctenophora

The last common ancestor of all living metazoans is thought to have split into two lineages that are around today: sponges and everything else (Figure 9.12). The "everything else" includes the Phylum Cnidaria and metazoans characterized by bilateral symmetry, grouped together in Bilateria. Cnidarians and bilaterians share some derived features that distinguish them from sponges. First, besides having specialized cells, cnidarians and bilaterians have **tissues**, essentially collections of specialized cells integrated into similar ECM. Cnidarians and bilaterians also go through a stage of development called **gastrulation** (Figure 9.14). Gastrulation occurs after the blastula stage and results in a stage called a **gastrula**.

In its simplest form, gastrulation involves pushing part of the wall of the blastula inward, similar to how you might push in part of an underinflated basketball. As a result, some of the cells that used to be on the outside of the

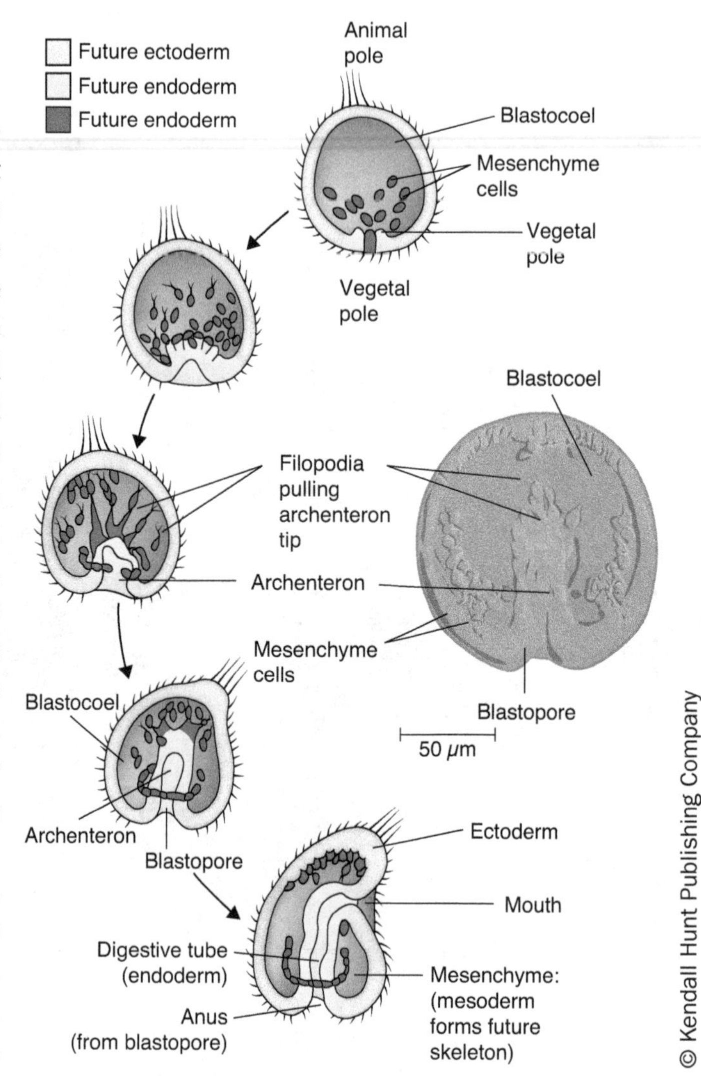

Figure 9.14. Gastrulation.

blastula are now on the inside, forming a second, internal layer of cells to go with the remaining outer layer. The inner layer is called the **endoderm** and the outer layer is called the **ectoderm**. These layers later produce different parts of the organism and are referred to as **germ layers**. In addition to a new layer of cells, there is now also a new space inside of the embryo, called the **archenteron**, which is continuous with the outside through an opening called the **blastopore**. The archenteron and the endoderm surrounding it are essentially the embryonic gut and typically form the digestive system in the adult.

Cnidarians (Figure 9.15) include some familiar metazoans: jellyfish, sea anemones, and hydra, among others. They also include corals, the tiny metazoans that build massive reefs from the shells they secrete around themselves. Cnidarians have a unique type of cell, a **cnidocyte**, which contains a specialized organelle called a **nematocyst**. Nematocysts propel a barb into prey that trigger the cell, delivering a toxin. Cnidarians are generally carnivores, using the cnidocytes on their tentacles to kill or immobilize prey, but some cnidarians, especially corals, also have symbiotic algae, called **zooxanthellae**, that provide nutrition to their cnidarian hosts.

The relationship between the gastrula and an adult cnidarian is easy to see. The cnidarian body consists of two layers, an outer and an inner, with a jelly-like substance called **mesoglea** in between. The archenteron becomes a sac-like gut, and the blastopore is ringed with tentacles and becomes the opening through which food is ingested and waste is expelled. Because they

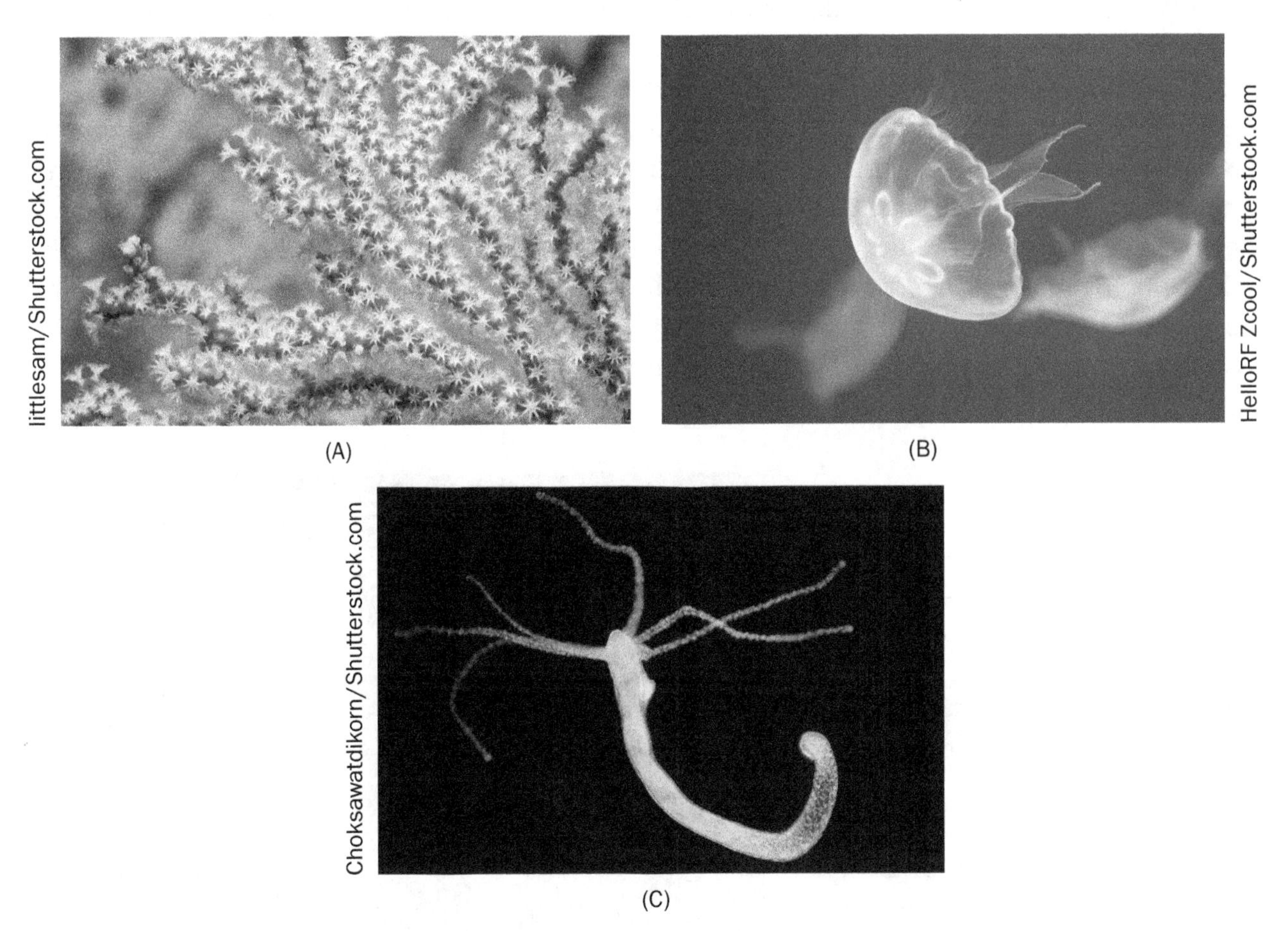

Figure 9.15. Phylum Cnidaria. (A) A sea fan coral, with the individuals extending their tentacles. (B) A jellyfish. (C) A hydra.

have two germ layers, cnidarians are referred to as **diploblasts**. Cnidarians have **radial symmetry**, meaning they are circular without an identifiable left or right side. Another way to describe radial symmetry is that you can cut a radially symmetrical object like a pie: there are multiple ways you could cut it in half and the two halves would be mirror images of each other. Besides the lining of the gut and the outer "skin," cnidarian tissues include nervous tissue and muscle.

Cnidarians exhibit alternation of generations, but, unlike plants, they do not alternate between diploid and haploid individuals. Instead, they alternate between asexually reproducing **polyps** and sexually reproducing **medusae** (s. medusa). Polyps are usually sessile and anchor themselves to a substrate, so their tentacles can extend into the water column and grab passing prey. Medusae, so called because the tentacles ringing the disk of the body bring to mind the snake-haired medusa of Greek mythology, are free-swimming and propel themselves by squeezing jets of water through the mouth-anus. Different groups of cnidarians can be distinguished in part by whether the polyp or medusa is dominant in the life cycle. Corals spend most of their lives as polyps, whereas the medusa is dominant in jellyfish.

Like sponges, cnidarians are all aquatic and mostly marine. Only corals produce a skeleton of calcium carbonate, and a coral reef is composed of thousands of individual skeletons that have cemented themselves together.

There is another phylum that is superficially similar to Cnidaria, the **Phylum Ctenophora** (Figure 9.16). Ctenophores are called **comb jellies**, because they have bands of cilia that resemble combs running along their transparent bodies, and because they superficially resemble jelly fish. Their bodies consist of a main bell or globe from which tentacles extend. Like jellyfish, comb jellies are marine, but unlike jellyfish they are filter feeders. While they are radially symmetrical, they have an important anatomical difference from cnidarians: they have a separate mouth and anus. Thus, food enters one opening, and waste exits by another at the other end of the digestive tube. Comb jellies also have nervous tissue, like cnidarians and bilaterians.

The phylogenetic relationships of comb jellies are still debated. Some studies have placed them at the very base of the metazoan tree, which would imply multiple origins of nervous tissue and a gut with two openings, but the current consensus places them with Bilateria.

9.6.3 Bilateria

The vast majority of metazoans are bilaterians, so called because they all exhibit bilateral symmetry, at least at some point in their lives. Bilaterally symmetrical objects have identifiable right and left sides; there is only one way to divide such an object—through the plane between the left and right sides—and get mirror image halves (Figure 9.17). Like cnidarians, bilaterians have tissues, but they also

Kondratuk Aleksei/Shutterstock.com

Figure 9.16. A comb jelly (Phylum Ctenophora).

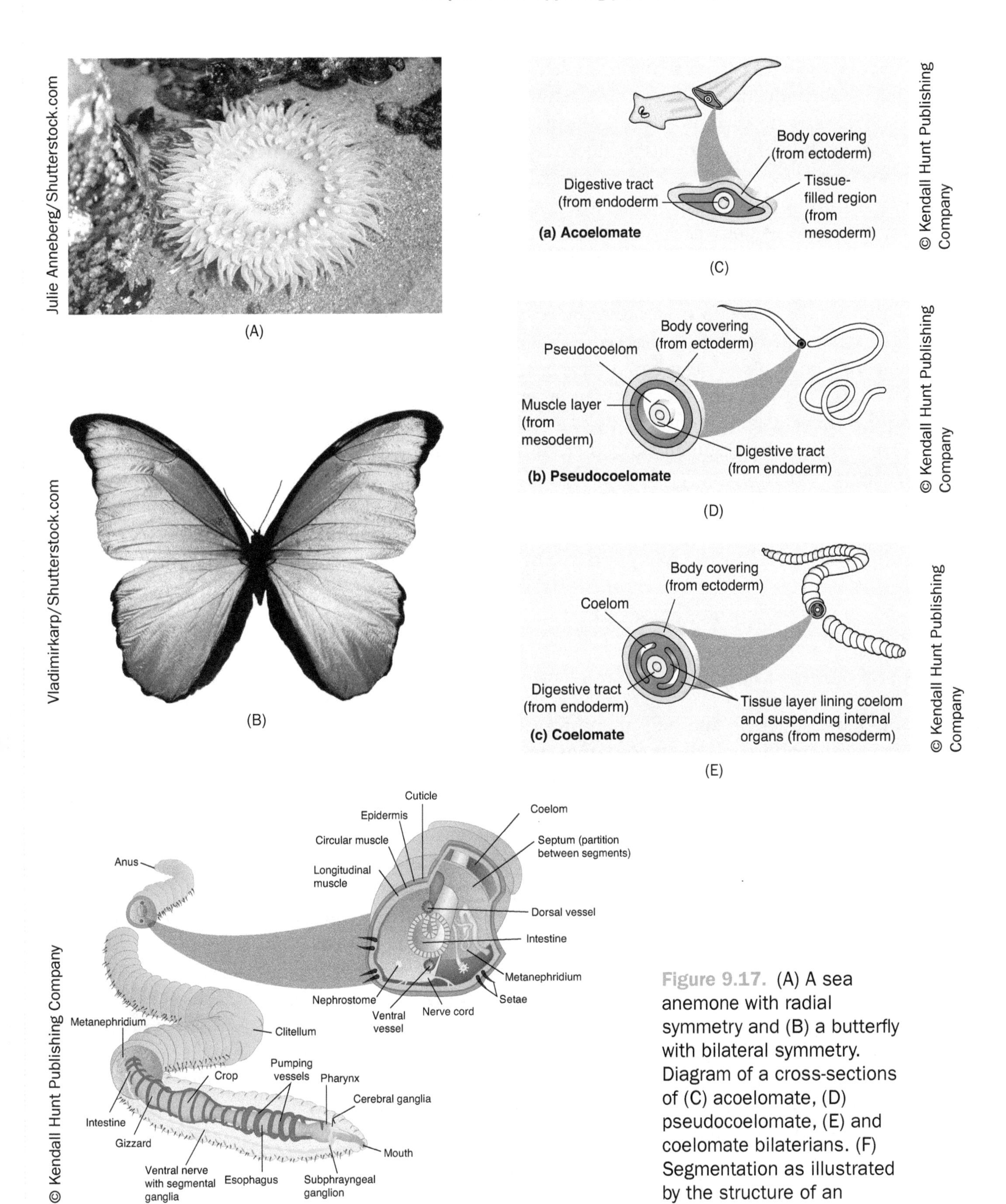

Figure 9.17. (A) A sea anemone with radial symmetry and (B) a butterfly with bilateral symmetry. Diagram of a cross-sections of (C) acoelomate, (D) pseudocoelomate, (E) and coelomate bilaterians. (F) Segmentation as illustrated by the structure of an earthworm.

have **organs**, specialized structures with specific functions consisting of multiple tissues. They also usually have a gut with a separate mouth and anus.

Bilaterians are **triploblasts**: they have three germ layers. In addition to the ectoderm and endoderm, bilaterian gastrulas produce a layer between these two called **mesoderm**. Mesoderm forms from some of the cells that are moving inwards during gastrulation. Bilaterians often have a **body cavity**, a fluid-filled internal space into which the organs extend. The most common type of body cavity is called a **coelom**, and it forms as a space within the mesoderm; such bilaterians are referred to as **coelomate**. Some bilaterians have a body cavity that forms between the endoderm and mesoderm; such bilaterians are called **pseudocoelomate**. Bilaterians that have no spaces between or within the germ layers are called **acoelomate**.

Another phenomenon observed in bilaterians is **metamerism** or segmentation. This refers to a body plan that is based on repeated units that contain the same set of structures. An interesting consequence of metamerism is that different groups can evolve by modifying the same basic segmented body plan in different ways, such as changing the number of segments, by fusing segments together, or by specializing a segment or set of segments for a particular function.

In the twentieth century, our picture of bilaterian phylogeny was based primarily on embryology anatomical patterns such as body cavities and segmentation. More recently, analysis of DNA sequences has given us a new picture of bilaterian phylogeny, some of which accords with the findings of earlier studies, but some of which came to very different conclusions about relationships among phyla.

9.6.4 Protostomes and deuterostomes

One point on which anatomical and molecular studies agree is the division of bilaterians into **protostomes** and **deuterostomes**, and these are reflected in the names of two multi-phylum groups, **Protostomia** and **Deuterostomia**. These two names were originally based on the fate of the blastopore in these groups. Bilaterians have a gut with two openings, one of which started as the blastopore, with the second forming later. In protostomes (which comes from the Greek for "mouth first"), the blastopore becomes the mouth, whereas in deuterostomes ("mouth second") the blastopore becomes the anus (Figure 9.18).

Protostomes and deuterostomes differ in two other embryological features. One has to do with how their cells divide before even the blastula stage, during what is known as **cleavage**. Protostomes have cleavage that is **spiral** and **determinate**. Spiral cleavage involves the cells dividing such that they end up overlapping each other, like bricks. Determinate cleavage refers to the fact that early on each cell is fated to become a particular portion of the organism. Thus, if you remove a particular cell from a protostome embryo at, say, the eight-cell stage, that embryo will be missing any structures normally derived from that cell. Deuterostomes have **radial cleavage** that is **indeterminate**. The cells after cleavage lie directly on top of each other, the same way the sections of an orange might be arranged if you cut it into halves, quarters, and then eighths, and so on. And if you remove a cell from the embryo during cleavage, you will still get a complete deuterostome developing from it (though it might be smaller than normal).

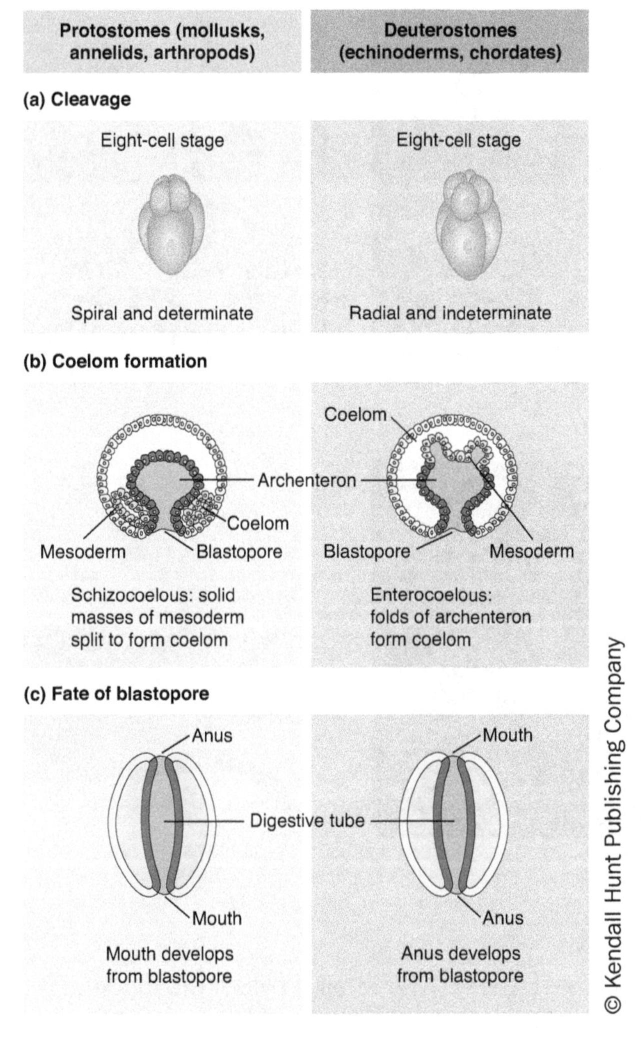

Figure 9.18. Differences between protostomes and deuterostomes.

The other feature that distinguishes protostomes and deuterostomes is how the coelom forms (at least in those members that have a coelom). Protostomes exhibit **schizocoely**, which means that the coelom forms by the initially solid mass of mesoderm internally pulling layers of cells apart from each other to form the space for the coelom within the mesoderm. Deuterostomes exhibit **enterocoely**, where the mesoderm forms as pockets pinching off the developing endoderm during gastrulation, and as the pockets pinch off of the endoderm, a space is retained inside of the mesoderm that becomes the coelom.

Protostomes include a number of familiar phyla (Figure 9.19): **Phylum Mollusca**, including **cephalopods** like squid and octopus, **bivalves** like clams and oysters, and **gastropods** like snails and slugs; **Phylum Annelida**, the segmented worms, including earthworms, leeches, and marine tube worms; and **Phylum Arthropoda**, including **insects** like beetles and ants, **arachnids** like spiders and scorpions, **crustaceans** like crabs and lobsters, and **myriapods** like centipedes and millipedes. All of these phyla are coelomate. One interesting

Figure 9.19. Examples of protostomes. (A) A snail (Phylum Mollusca). (B) An earthworm (Phylum Annelida). (C) A roundworm (Phylum Nematoda; stained to show internal structures). (D) A crab (Phylum Arthropoda).

result from molecular data is the placement of a couple of noncoelomate phyla. **Phylum Nematoda** includes the roundworms, which are pseudocoelomate but are considered to be closely related to arthropods. The acoelomate flatworms in the **Phylum Platyhelminthes** (including the planarians observed in many biology labs) are placed with mollusks and annelids. Protostomes are found all over the world in almost every environment, and they include the most diverse group of metazoans, the insects.

Deuterostomes include two familiar phyla: **Phylum Echinodermata** and **Phylum Chordata** (Figure 9.20). Echinoderms are marine and include sea stars, sea urchins, sand dollars, and sea cucumbers, among others. Adult echinoderms have a **pentaradiate symmetry**, which means they have a sort of radial symmetry that divides their bodies into five portions, or into portions that are multiples of five, like the symmetry of a five-pointed star. While adult echinoderms do not appear to be bilaterally symmetrical, their larvae clearly are.

Figure 9.20. Examples of deuterostomes. (A) A sea star (Phylum Echinodermata). (B) A sea squirt (Phylum Chordata). (C) A lancelet (Phylum Chordata).

Chordates (Figure 9.20) include three subphyla: **cephalochordates**, including lancelets; **urochordates**, including sea squirts and tunicates; and **vertebrates**. All chordates share four features that are present at some point in the development of the chordate: a flexible rod of connective tissue along the back called a **notochord**; **slits** in the throat, or **pharynx**, that in some cases become **gills**; a **single, dorsal hollow nerve cord**, meaning it has a space in the middle and runs on top of the notochord; and a **tail** that extends beyond the anus. Cephalochordates and urochordates are marine filter feeders that rely on cilia to create currents for feeding, but vertebrates are active feeders relying on muscles.

9.7 Vertebrates

We pay special attention to vertebrates because we are vertebrates, but vertebrates also have a well-developed internal skeleton that has a complex anatomy and that has left a rich fossil record. The earliest vertebrates are known from the middle of the Cambrian, over 500 million years ago. All vertebrates have a **skull** and some development of **vertebrae** and a **vertebral column** that forms their spine, incorporating or replacing the notochord. Figure 9.21 shows a phylogeny of vertebrates. The first vertebrates were aquatic and did not come onto land until about 380 million years ago. The ancestry of terrestrial vertebrates is found among those aquatic vertebrates commonly called fishes.

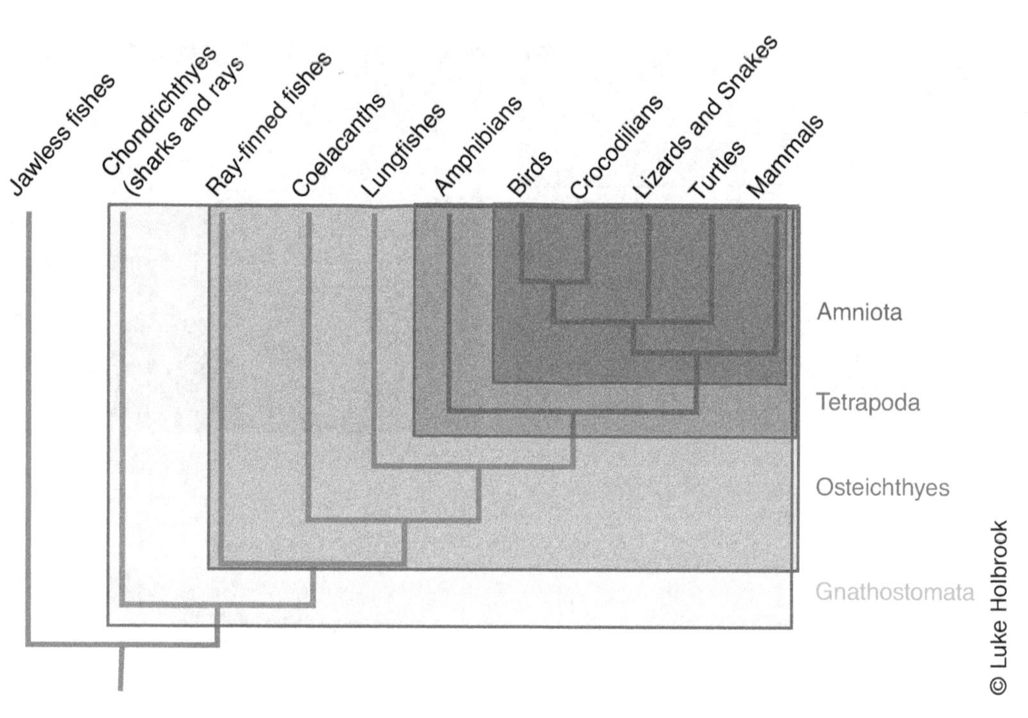

Figure 9.21. A phylogeny of living vertebrates.

9.7.1 Fish diversity

Most vertebrates are fishes (Figure 9.22), which is used as a generic term for nonterrestrial vertebrates and their descendants. The earliest vertebrates were **jawless fishes**, of which today there are two groups still present: **hagfish** and **lampreys**. As mentioned in Chapter 5, hagfish do not even have vertebrae as adults, but it appears that they have lost vertebrae secondarily, and their closest living relatives are the lampreys, which have small but distinct vertebrae. Hagfish are scavengers of the deep ocean, whereas lampreys attach themselves to other fish and feed on their blood.

Most vertebrates have **jaws**, and jawed vertebrates are called **gnathostomes**, as they comprise a monophyletic group. Jawed vertebrates share some other features, including paired appendages (fins or limbs). Jawed vertebrates can be divided into two living lineages: **Chrondrichthyes** ("cartilaginous fishes") and **Osteichthyes** ("bony fishes"). Chondrichthyes include sharks, skates, and rays, as well as an unusual deepwater group called ratfishes. Chondrichthyans typically have a skeleton of cartilage, which is why, although there are lots of fossils of chondrichthyans, the vast majority of these are teeth, as the cartilage skeleton decays quickly.

The term Osteichthyes was originally applied only to fishes with skeletons of **true bone**, that is bone that has cells in its matrix, like our bones. But some bony fishes are more closely related to terrestrial vertebrates (which also have true bone) than they are to other bony fishes. For that reason, terrestrial vertebrates are now included in Osteichthyes—essentially

Figure 9.22. Fishes. (A) The mouth of a lamprey, a jawless fish. (B) Hammerhead sharks, a type of cartilaginous fish (Chondrichthyes). (C) A school of tuna, a type of ray-finned fish (Actinopterygii). (D) A specimen of a coelacanth, a lobe-finned fish. (E) A lungfish, another lobe-finned fish.

making us "bony fishes." While at first it might seem silly to say that we are fishes, this just acknowledges that our ancestry lies with fishes.

Osteichthyes is divided into two monophyletic groups: **Actinopterygii** (ray-finned fishes) and **Sarcopterygii** (lobe-finned fishes). Actinopterygians are the most diverse group of vertebrates, including about 20,000 species. Most of the creatures that you think of when you hear the word "fish" are actinopterygians; moreover, they are most likely part of a specific group of actinopterygians, **teleosts**. With few exceptions, any fish that you eat, that you catch in a lake, stream, or ocean, or that you keep in an aquarium is a teleost.

The term Sarcopterygii has a history analogous to that of Osteichthyes. Originally it was coined for fishes that have more limb-like fins and that share certain features with terrestrial vertebrates, but it came to include terrestrial vertebrates because some lobe-finned fishes are more closely related to terrestrial vertebrates than to other lobe-finned fishes. So, besides being bony fishes, we are also lobe-finned fishes. There are two kinds of living lobe-finned fishes. **Coelacanths** were originally thought to have gone extinct with the dinosaurs, until one showed up in a South African fish market. The first population of these fishes was discovered along the Comoros Islands off the east coast of Africa, and populations have since been discovered in the waters off of Indonesia and off of South Africa. The other lobe-finned fishes are **lungfishes**, which are found today in Africa, South America, and Australia. They live in freshwater, often in places where the waters dry up seasonally. The South American and African lungfish can burrow into the mud and **aestivate**, meaning they go into a dormant state until the rains return and fill up their streams and ponds. Our closest living fish relatives are lungfish, which do have lungs that they use to breathe air. However, lungs are actually also found in some ray-finned fishes that use them to breathe air in low-oxygen freshwater environments, and it appears that the origins of lungs go back to the ancestry of Osteichthyes.

9.7.2 Tetrapods and the transition to land

We have already made several references to terrestrial vertebrates, which are grouped together in the **Tetrapoda**, a name that means "four feet" (Figure 9.23). Of course, not all tetrapods are terrestrial, as some lineages, like whales and dolphins, have returned to the water. Likewise, not all tetrapods have four limbs, the best example being snakes. Both whales and snakes are tetrapods because they share a recent common ancestry with other tetrapods. Regardless, **four limbs with digits** is a characteristic feature of tetrapods as a whole. The earliest tetrapods were still aquatic, but they had limbs with digits. In fact, although living tetrapods have five or fewer digits on each limb, the earliest tetrapods had as many as eight.

Tetrapods are divided into two living lineages, **Amphibia** and **Amniota**. Amphibians include familiar creatures like frogs and salamanders, as well as a strange group of legless amphibians called caecilians. Amphibians typically have an aquatic larval stage, such as a tadpole, that develops into a terrestrial adult. Fertilization of eggs and the development of fertilized eggs into larvae usually requires a moist environment, if not a body of water, so amphibian reproduction is often tied to water. Amphibian skin is not very waterproof and is permeable to many substances. In fact, many amphibians "breathe" through their skin, exchanging oxygen and carbon dioxide with the environment through blood vessels near the surface of the skin. This permeable skin has its drawbacks: it makes it easier to lose water, which is why few amphibians live in deserts; no amphibians can live in marine environments, where the salt water would actually dehydrate them through osmosis; and amphibians are some of the most sensitive organisms to human-produced chemicals that pollute the environment, because these chemicals have an easier time getting into their bodies.

Figure 9.23. Examples of tetrapods. (A) A salamander, an amphibian. (B) A turtle. (C) A snake. (D) A shoebill, a kind of bird. (E) A young crocodile. (F) A tree shrew, a kind of mammal.

The features of **amniotes** allow them to avoid some of the issues that amphibians face. Amniotes have **waterproof skin**, due to an abundance of the protein **keratin** in dead cells in the outermost layer of the skin. Amniotes get their name from the **amniote** egg. Besides often having a hard egg shell to prevent water loss, the embryos of amniotes are surrounded by four **extraembryonic membranes**, including the **amnion** that gives the amniote egg its name. These fluid-filled membranes effectively create a self-contained moist environment, so that amniotes do not need to lay their eggs in water. Amniotes have **internal fertilization**, where sperm are deposited inside the female, so they are not reliant on water in the environment for fertilization.

Amniotes include **birds**, **mammals**, **crocodiles**, **lepidosaurs** (lizards and snakes and the tuatara of New Zealand), and **turtles**. Back in Chapter 5, we noted that the last three groups in that list, which we traditionally call "reptiles," are not a monophyletic group, so either birds are included in **Reptilia**, or we give the bird–"reptile" lineage another name, **Sauropsida**. Amniotes can be divided into two major lineages: Sauropsida (or Reptilia) and **Synapsida**. Synapsids include mammals and their extinct relatives.

9.8 Mammals

Because we are mammals, we pay particular attention to mammal diversity and evolution. Mammals first appear in the Mesozoic, but it is not until the extinction of the dinosaurs that mammals start to occupy the diversity of niches that we associate with them today. The first mammals were small and adapted for eating insects. It was thought that mammals stayed that way until the dinosaur extinction, but a series of discoveries have shown that mammals got as big as beavers and included climbing, digging, swimming, and even gliding forms. Still, Mesozoic mammals were generally smaller and less ecologically diverse than Cenozoic mammals.

Mammals are distinguished by a number of features, notably having fur and nursing their young with milk. Mammals, like birds, have **endothermy**, which means that they maintain a high body temperature by having their cells produce heat. **Ectotherms**, like most other vertebrates, do not produce enough heat from their cells to maintain a body temperature high enough for their metabolic processes, so they are dependent on the temperature of the environment to maintain a metabolism sufficient for activity. As endotherms, mammals are able to remain active in cooler climates, such as at higher latitudes and at higher altitudes, and they can remain active at night. Mammals are generally more active than ectothermic tetrapods, and they have a more upright stance, with the legs more directly under the body.

Endothermy does have a steep cost: in order to keep cells "running" at a high enough rate to produce heat, mammals need a lot of food and oxygen. Mammals eat far more often than ectotherms, and most mammals have teeth specialized for **chewing**, rather than just holding onto prey. Chewing processes the food before it gets to the chemical digestion in the stomach. Mammals also breathe more often than ectotherms, and they have features for powering that breathing, for minimizing loss of water and heat when they exhale, and for allowing them to breathe while they are chewing.

9.8.1 Monotremes and marsupials

Living mammals can be divided into three lineages (Figure 9.24): **Monotremata**, **Marsupialia** (also called **Metatheria**), and **Placentalia** (also called **Eutheria**). Monotremes are mostly noted for their ancestral features. They are the only egg-laying mammals, they have no nipples for their young to suckle, and they have skeletal similarities to fossil nonmammal synapsids. Yet, the two types of living monotremes, the platypus and the echidna or spiny anteater, are by no means primitive mammals. In fact, they are quite unusually specialized. The platypus has beaver-like adaptations for a semiaquatic life and has a unique, sensitive duck-like "bill" that helps it find invertebrates in the sediment at the bottoms of the streams where it lives. The echidna has many of the adaptations for ant-eating that we mentioned in Chapter 5—a long toothless snout, a long sticky tongue, and powerful forelimbs with heavy claws for digging into the nests of ants and termites. It is also covered in sharp quills, modified hairs that protect it from predators. Monotremes today are found only in Australia and New Guinea.

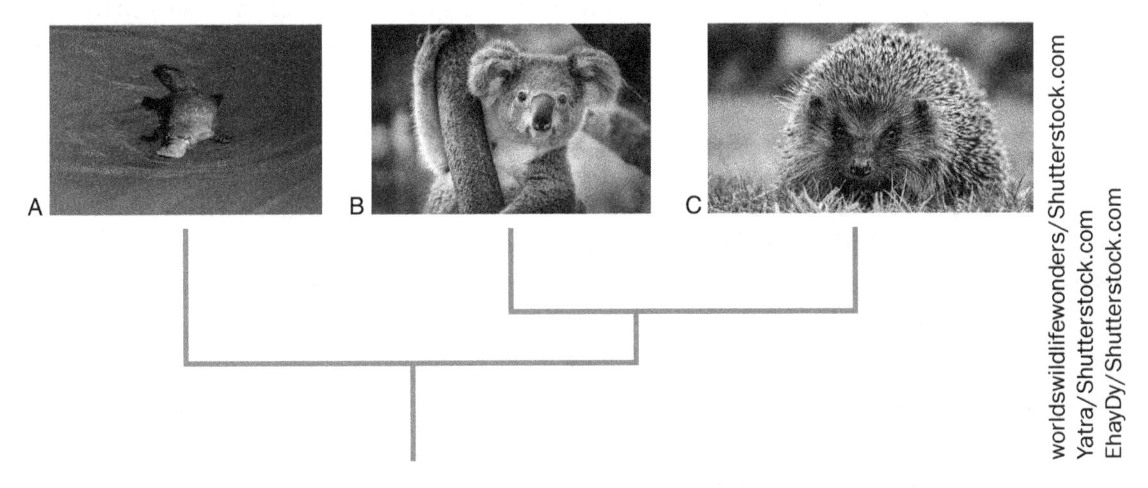

Figure 9.24. The three major lineages of mammals and their relations to one another. (A) A platypus. (B) A koala. (C) A hedgehog.

Marsupials and placental mammals are united in a monophyletic group called **Theria**, characterized by live birth, nipples for nursing young, and certain details of the skeleton. Most marsupials are found today in two regions, Central and South America and Australia and New Guinea. Most of marsupial diversity is in Australia, including the most familiar marsupials, like kangaroos and koalas. South American marsupials include a variety of opossums, including relatives of the only North American marsupial, the opossum (*Didelphis virginiana*). The word "marsupial" comes from the Greek word for pouch and refers to the **pouch** in which females of many species carry their young. However, not all marsupial species have such pouches.

Despite the implication of marsupials not being placental mammals, marsupials actually do have a **placenta**, an organ that combines tissues of the embryo and the mother to facilitate exchange of nutrients and wastes between the two. The main difference between marsupial and placental reproduction is the relative time that developing young spend connected to the mother by the placenta versus the amount of time they spend nursing. To put it another way, marsupials and placentals differ in how much of development is funded by exchange through the placenta rather than by milk. Developing marsupial young spend a relatively short amount of time in the mother's womb, and this is reflected by the degree of development—or, from our perspective, underdevelopment—the young exhibit when they are born. Newborn marsupials are small, hairless, and lack hindlimbs. Their best developed features are their forelimbs (for crawling from the womb to the pouch) and their jaws (for latching onto the nipple). Once they attach to a nipple, they continue to develop until they essentially have all of the features of a small adult and can leave the pouch for good.

9.8.2 Placental mammals

In contrast to marsupials, placental mammals spend relatively more time of development attached to the placenta and are born with essentially all of the anatomy that they will have as adults. During nursing, development is mainly about growth and changes in proportions. Placental mammals are far more diverse than marsupials and are the dominant native mammals populating all of the continents except marsupial-dominated Australia.

Placental mammal diversity spans a wide variety of ecological roles, body sizes, diets, and styles of locomotion. Today, the largest metazoans on land and in the sea are placental mammals, but the most diverse placental mammals are small species; the largest order of mammals by far is the **Order Rodentia**, the rodents. The second most diverse order is the **Order Chiroptera**, the bats, which are also one of only three groups of vertebrates that have achieved powered flight, the others being birds and the extinct pterosaurs. Placental mammals have also evolved multiple groups of large terrestrial herbivores, as well as carnivores that prey upon them. There are also multiple lineages of mammals that have evolved an aquatic lifestyle and returned to the water, including whales and dolphins; manatees and dugongs; and seals, sea lions, and walruses.

9.8.3 Primates

The mammals that are of greatest interest to humans are those in the **Order Primates** (Figure 9.25), because that is the order to which we belong. The features that are unique to primates are mostly technical details of their skeletal anatomy. Primates tend to have larger brains than other mammals, **opposable thumbs** that make gripping easier, and they have **stereoscopic vision**, where the images formed in each eye go to both halves of the brain, allowing us to see more easily in three dimensions. None of these features is unique to primates, but they do point to an **arboreal** ancestry—that is to say that primate ancestors were tree-dwellers.

The relationships among the major lineages of primates that morphologists and paleontologists had inferred over the previous century have been confirmed by analyses of primate DNA sequences. Primates are divided into two main branches: **Strepsirhini** ("split nose") and **Haplorhini** ("simple nose"). Strepsirhines include the lemurs of Madagascar and the lorises and galagos of Africa and Southeast Asia. The name refers to their snouts, which

Figure 9.25. Primates. (A) A lemur. (B) A tarsier. (C) A squirrel monkey, a New World monkey. (D) A baboon, an Old World monkey. (E) A gibbon, a lesser ape. (F) An orangutan, a great ape.

are longer and "split" at the bottom like that of many mammals, such as dogs and cats. Haplorhines lack this "split" nose and include what we informally call monkeys and apes, as well as the small, bug-eyed **tarsier** (Genus *Tarsius*) of the islands of Indonesia and Sulawesi.

Monkeys and apes are included in Anthropoidea, which is split into **Platyrrhini** ("flat nose") and **Catarrhini** ("down nose"). Platyrrhines include the **New World monkeys**, such as howler monkeys, spider monkeys, squirrel monkeys, and marmosets, which are native to

Central and South America. Catarrhines include apes and **Old World monkeys**. Because Old World monkeys are more closely related to apes than they are to New World monkeys, there is no group in our classification that corresponds to "monkeys." Old World monkeys are a monophyletic group (Family Cercopithecidae) found today in Africa and Asia that includes macaques, baboons, and leaf-eating colobus monkeys, among others.

Apes include gibbons, orangutans, gorillas, chimpanzees, and humans, and are classified together in a superfamily, **Hominoidea**. Apes are split into the **lesser apes**, including only the **gibbons** (Family Hylobatidae), and the **great apes**, which are in the **Family Hominidae**. In the past, Hominidae was reserved for humans and their extinct relatives, and other great apes were placed in the Family Pongidae. But this was done to preserve the notion that humans were distinct from other animals, and it meant that Pongidae was not monophyletic. We include humans in the same family as the great apes to make it monophyletic, and because the name Hominidae was established before the name Pongidae, Hominidae has become the family name for all great apes. Today, apes are found in two places: Southeast Asia (gibbons and orangutans) and Africa (gorillas and chimpanzees).

9.9 Human origins and evolution

We will finish our survey with a brief review of the evidence for the origins of our species. In the late nineteenth and early twentieth centuries, there were a number of fossil discoveries that were thought to establish the birthplace of the human species on different continents, including Europe, Asia, and even one argument for North America based on what turned out to be the tooth of a fossil pig. Figure 9.25 indicates that our closest relatives are African apes, and it was twentieth-century discoveries of fossil human relatives in Africa that established that continent as the cradle of human evolution.

9.9.1 Divergence from African apes

Based on both molecular clocks and recent fossil discoveries, the lineages leading to chimpanzees (or African apes in general) and to humans diverged from their last common ancestor about seven million years ago. We continue to find fossils that fill out the seven-million-year history of the human lineage, and these have helped to tease out the sequence of evolutionary changes that make us distinct from our closest living relatives (Figure 9.26).

We differ from chimpanzees in several respects, but we can group our differences into two suites of anatomical features. First, humans have a number of anatomical features related to our unique form of **bipedalism**, or walking on two legs. Bipedalism occurs or occurred in other groups, notably birds and dinosaurs, but also kangaroos and even some extinct crocodilians. But humans are unique in how they hold their bodies upright, rather than keeping the vertebral column in the same horizontal orientation as a quadrupedal tetrapod, the way kangaroos, birds, and other bipeds do. As a result, our **vertebral column is arranged differently** to bear weight vertically (leading to our unique suite of back ailments) and the **shape of the pelvis** is different. We also have unusual feet, with long soles and an **enlarged first toe** (the big toe) that is the main point of contact with the ground when we push off during a step.

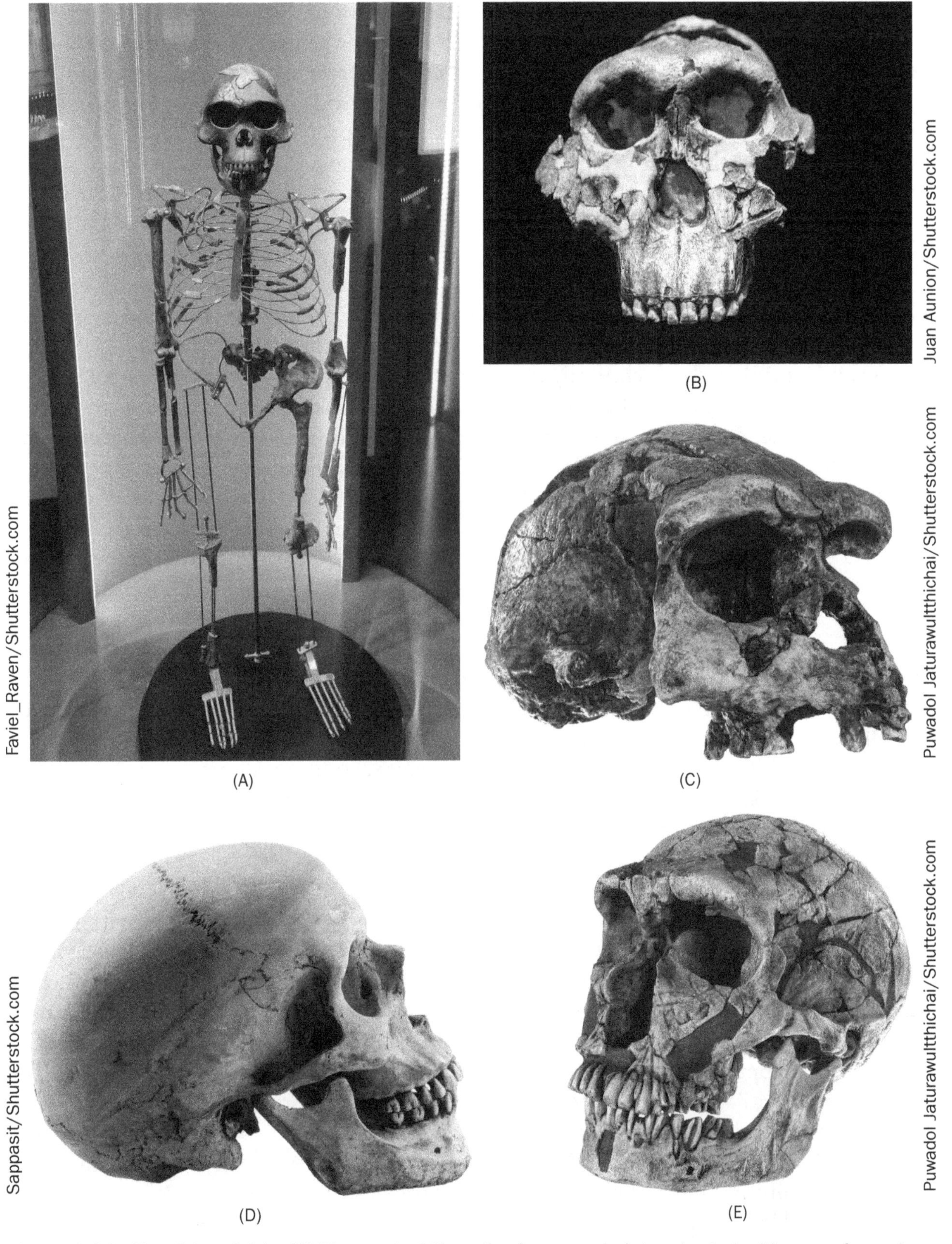

Figure 9.26. Fossil hominids. (A) The cast of "Lucy," a famous skeleton *Australopithecus afarensis*, an "australopithecine." (B) A skull of *Paranthropus boisei*, an "australopithecine." (C) A skull of *Homo erectus*. (D) A skull of a modern human, *Homo sapiens*. (E) A skull of *Homo neanderthalensis*.

The other set of uniquely human features relates to our heads. Even for primates, humans have unusually **large brains** relative to their body size, and the part of the skull that houses the brain is enlarged and rounded, without the **sagittal crest** that we see on the skulls of many mammals, including many primates, and which provides attachment for strong jaw muscles. Humans also have **short faces**; our jaws do not protrude as they do in other primates (a condition called **prognathism**), and our lower jaws have a distinct **chin**, which helps to buttress the front of the jaw.

9.9.2 Fossil hominins

The fossils that illustrate evolution of the lineage unique to humans are sometimes referred to as fossil hominids, but technically "hominid" refers to all great apes. Here we will use the term **fossil hominins**; this technically is also not quite correct, but it at least is not a term that would be confused with others we have used in this chapter.

The earliest fossil hominins are found in Africa, which is consistent with the close relationship between humans and African apes. In fact, Africa, as we will see, is where our species first arose and evolved until it started to spread around the globe. We now know of a whole series of early hominins from Africa that we will broadly call "**australopithecines**." In fact, these taxa do not form a monophyletic group, but rather some are more closely related to us than they are to each other. We can still make some general statements about what these taxa are like and what they tell us about the sequence of human evolution.

Australopithecines are remarkable for exhibiting a combination of ape-like traits and human-like traits. To varying degrees depending on how closely related they are to humans, australopithecines exhibit features related to bipedalism: vertical arrangement of vertebrae, human-like pelvis, and a first toe that is in line with the others, rather than an opposable first toe like apes have. Australopithecine skulls are still quite ape-like: prognathic jaws without a chin, skulls with relatively small braincases, thick brow ridges, and a prominent sagittal crest.

Australopithecines—in the broad sense used here—were in Africa from about 5.5 million years ago until about 1.2 million years ago. Our genus appears over two million years ago in Africa in the form of the earliest species of *Homo*, *Homo habilis*. The genus *Homo* is distinguished from australopithecines by a larger, rounder braincase with little to no sagittal crest. Early species like *Homo habilis* and *Homo erectus* retain thick brow ridges and prognathic jaws. Tool use is also associated with the genus *Homo*, although it is difficult to be certain that australopithecines did not also use tools. The earliest evidence of controlling fire is associated with *Homo erectus*.

Homo erectus appears a little later than *Homo habilis*, about 1.8 million years ago, but the species persists until as recently as 140,000 years ago. Our own species, *Homo sapiens*, is first known from fossils in Africa by 160,000 years ago and from tools attributed to the species as early as 200,000 years ago. Thus, *Homo habilis* and *Homo erectus* were alive at the same time as some australopithecines and likely lived in close proximity to them, and *Homo erectus* persisted well after *Homo sapiens* had appeared. *Homo erectus* is also remarkable for something else it shares with *Homo sapiens*: expanding its range out of Africa. Some of the earliest discovered fossils of *Homo erectus* were famously known as "Peking Man" and "Java Man," as they were discovered in China and Indonesia, respectively. Other specimens of *Homo erectus* have been discovered in the Middle East and in Europe.

Homo sapiens is distinguished from other species in the genus *Homo* by its even larger brain and its short face, weak brow ridges, and prominent chin. By 120,000 years ago, fossils of *Homo sapiens* are indistinguishable from the skeletons of modern humans. *Homo sapiens* starts spreading around the world around 100,000 years ago, reaching Asia and the Middle East by 70,000 years ago, Australia around 50,000 years ago, Europe around 40,000 years ago, and finally reaching the Americas about 15,000 years ago. Note that expansion into Australia means that humans were using watercraft by 50,000 years ago. About 50,000 years ago is also when humans exhibit a new level of innovation and culture: they make finer, more complex tools, and they produce art, such as various cave paintings known from this time.

We should mention one other species in our genus: *Homo neanderthalensis*, the Neanderthals. Once thought to be part of *Homo sapiens*, Neanderthals are now considered to be a distinct species. They are known from Europe and southwestern Asia and lived between 400,000 and 40,000 years ago. They therefore overlapped with *Homo sapiens*, though they soon go extinct shortly after *Homo sapiens* arrives in the areas where Neanderthals were present. Ancient DNA recovered from Neanderthals reveals that humans have some small percentage of Neanderthal genes, indicating that humans and Neanderthals hybridized in the past, but not enough to consider them to be the same species. Contrary to their depiction throughout most of the twentieth century, Neanderthals had a fairly complex culture comparable to their *Homo sapiens* contemporaries.

Further Reading

Brusca, R C., W. Moore, and S M. Schuster. 2016. *Invertebrates*. Sinauer Associates, Sunderland, MA.

Pough, F H., C M. Janis, and J B. Heiser. 2012. *Vertebrate Life*. Pearson Publishing, Upper Saddle River, NJ.

CPSIA information can be obtained
at www.ICGtesting.com
Printed in the USA
LVHW020752050119
602848LV00002B/3/P

9 781524 955656